The Problem of Information

An Introduction to Information Science

Douglas Raber

The Scarecrow Press, Inc.
Lanham, Maryland, and Oxford
2003

Z665 .R24 2003
Raber, Douglas, 1949-
Problem of information.

SCARECROW PRESS, INC.

Published in the United States of America
by Scarecrow Press, Inc.
A Member of the Rowman & Littlefield Publishing Group
4501 Forbes Boulevard, Suite 200, Lanham, MD 20706
www.scarecrowpress.com

PO Box 317
Oxford
OX2 9RU, UK

Copyright © 2003 by Douglas Raber

All rights reserved. No part of this publication may be reproduced, stored in a retrieval system, or transmitted in any form or by any means, electronic, mechanical, photocopying, recording, or otherwise, without the prior permission of the publisher.

British Library Cataloguing in Publication Information Available

Library of Congress Cataloging-in-Publication Data

Raber, Douglas, 1949–
The problem of information : an introduction to information science / Douglas Raber.
p. cm.
Includes bibliographical references and index.
ISBN 0-8108-4567-9 (alk. paper) — ISBN 0-8108-4568-7 (pbk. : alk. paper)
1. Information science. 2. Information theory. I. Title.
Z665 .R24 2003
020—dc21

2002030298

The paper used in this publication meets the minimum requirements of American National Standard for Information Sciences—Permanence of Paper for Printed Library Materials, ANSI/NISO Z39.48-1992.
Manufactured in the United States of America.

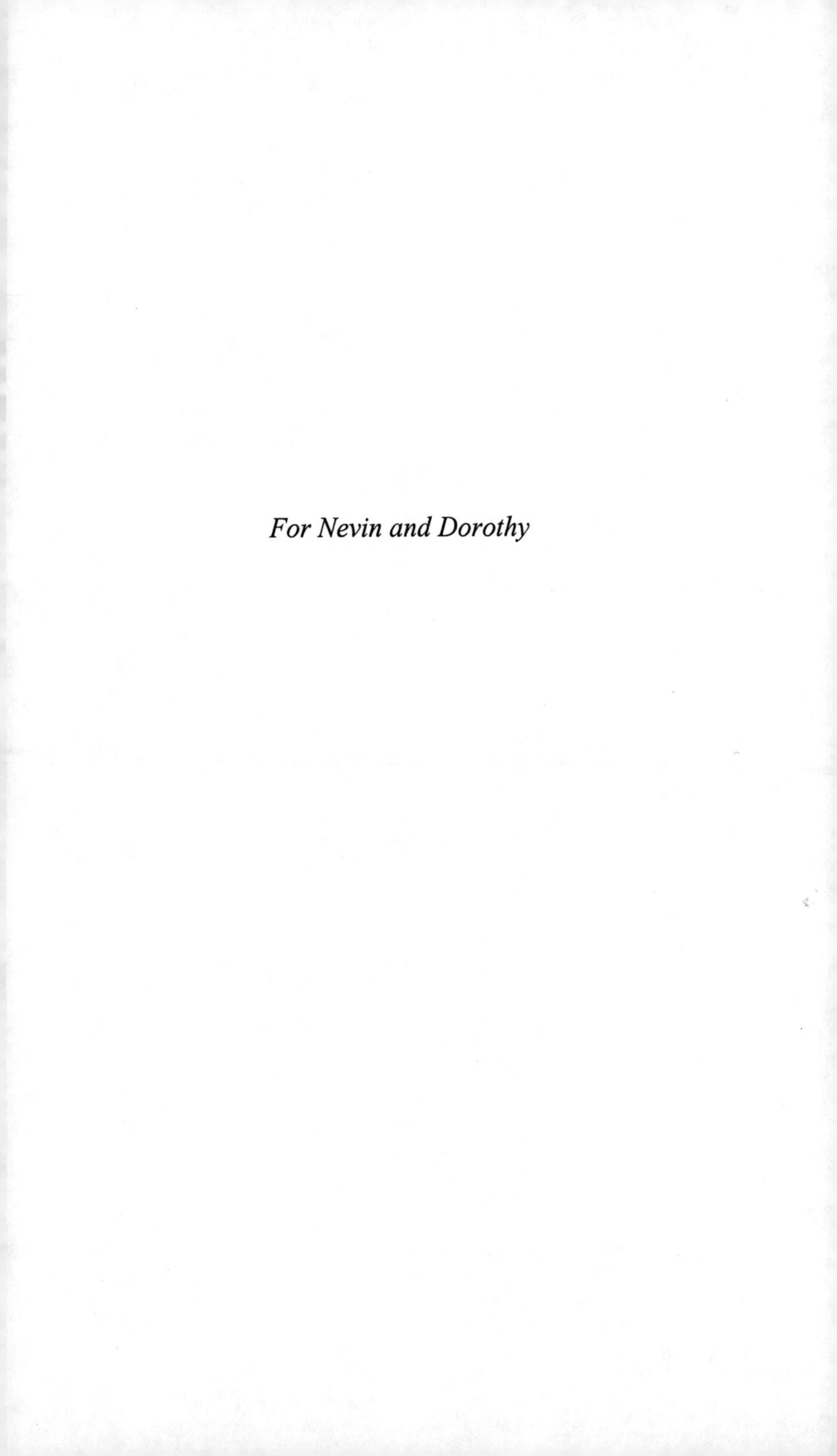

For Nevin and Dorothy

Contents

Preface

Most students in library and information science (LIS) programs do not have a background in information science. While many of them have experience in library or information work and intuitively recognize concepts presented in the research literature, the vocabulary of this research discourse is rarely understood completely. Similarly, there are many fine books available that offer an overview of the discipline but presume a familiarity with its discourse on the part of the reader.

Solutions to the challenges of managing information are equally elusive, as there are a variety of ways to think about and accomplish the organization and retrieval of information. For the most part, to do so depends on interpreting which particular challenges are to be solved. Paradoxically, however, the nature of information varies according to how it is organized, retrieved, and used.

This, then, is the "problem of information" proposed in the tile of this book; the protean nature of what Michael Buckland has called "information-as-thing," "information-as-process," and "information-as-knowledge." Over the course of this book, I will address the nature of information as a theoretical object, introduce the questions information science asks about information, and provide an explanation of how the discipline goes about doing that, all the while introducing the concepts of information science as if for the first time.

This book was written with beginning LIS students in mind; it should be accompanied by the reading of contemporary journal articles from the literature of information science. For those who intend further study of information science, additional material addressing methodological issues will be required. For budding librarians my goal is to make it clear that librarianship and information science are not as alien to one another as they sometimes appear to be.

Finally, I hope that this book may also serve as an introduction to others outside the fields of either librarianship or information science who are curious about the research issues and conceptual approaches of information science. Information science borrows heavily from other disciplines, especially the social sciences, but it adapts what it borrows. The resulting innovations may prove useful to the social sciences as these disciplines begin to confront issues of their own that grow from the human necessities and problems of an information age.

1

Information Science and the Problem of Information

Information science is not an easy discipline to describe to people who do not practice it, at least in part because the phrase itself is rather ambiguous. It is not entirely clear what is meant by "information" and how it might be rendered a "science." For example, if the term *information science* is used as a subject descriptor, conceivably it can be applied to such a wide variety of documents that any good indexer would be forced to doubt its effectiveness as an access point. Cybernetics, computer science, communications engineering, information technology, documentation, and librarianship can all be classified as information sciences, and there are many more besides.[1] What, then, distinguishes information science from the vast array of information sciences? The answer turns on the way the various information sciences conceive the nature of information. "Information" is a very accommodating concept. If we take the word to represent any object or phenomenon that might be informative, "then *everything* is, or might well be, information."[2]

In fact, one well known information scientist, Robert Fairthorne grew so frustrated that he pronounced that the word *information* be avoided if at all possible to prevent becoming downright inane or vacuous; he claimed that any use of the word might refer to some thing, event, or phenomenon that is somehow useful to someone.[3] In other words, the task of information science is to manage, maintain, and make available to various users the diversity of human discourse, as manifest in different sources and used for a multitude of purposes. Clearly, the challenge implied in so much diversity is considerable and eliminates from consideration a number of questions better addressed by disciplines that "intentionally neglect the semantic property of some information in order to derive more abstract definitions which can have broader applications."[4] The explicit examination of the semantic property of information, he continues, "implies that information science is specifically concerned with information *in the context of human communication*."[5] Information science, then, revolves around the communicative nature of information, the nature and purposes of its users, and the nature and purposes of its use.

At the very heart of this confusion lies the fact that "information" is a complex sign characterized by considerable ambiguity. Understanding "information" is less a matter of providing a formal definition of the word than apprehending what it means when it is used in a particular context. According to Ferdinand De Saussure, the sign is the fundamental unit of language. It is a double entity formed by the unity of a sound image or word and a concept that by convention and use is associated with a given sound image or word. Thus, a sign is a "two-sided psychological entity" whose "two elements are intimately united, and each recalls the other."[6] Saussure calls one of the these elements (the sound image) the "signifier," and the other (the concept) he calls the "signified." Jere Paul Surber puts it this way:

> According to Saussure, language consists of signs and their interrelations. For Saussure, a sign must always be understood (in a way rather different from our ordinary notion of this) as a relation between a signifier and a signified. A signifier, in Saussure's definition, is any sensory image that is immediately connected with (or gives rise to) a specific mental concept or idea; a signified is the mental concept or idea thus associated with a signifier. In reading Saussure, it is crucial to bear in mind that a sign is never identified exclusively with either the signifier or the signified; rather it is the relation between the two.[7]

In Figure 1.1, representing the sign, the arrows indicate that the relation between signifier and signified is both simultaneous and reciprocal. In this context, the "problem of information" becomes one of understanding the nature of the sign, 'information.' If the sound-image/word *information* is a signifier, then what does it signify?

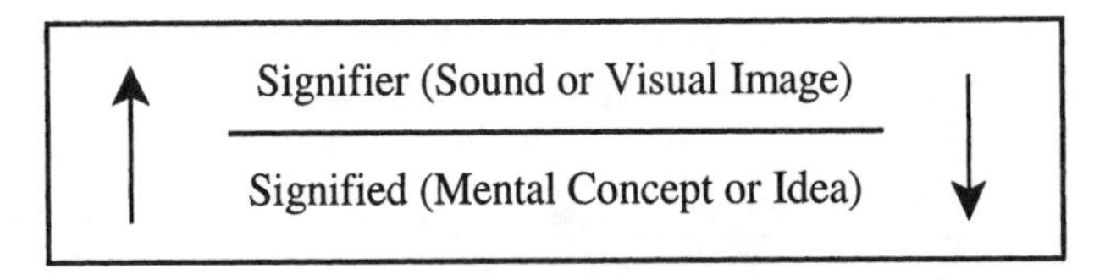

Figure 1.1 The Sign

Some signs give us relatively few problems. The cultural conventions and understandings that govern the relations between signifier and signified are strong, secure, and widely shared. Within the body of speakers of English, the sign "cat" almost immediately calls to mind the image of a small, four-legged, predatory animal, probably accompanied by a keen gaze and whiskers. Among American drivers, a red octagonal sign with the word *STOP* printed in bold white letters on the surface facing a driver is likely to be immediately understood. Neither sign, however, is quite as simple as it seems. "Cat" may require clarification regarding the speaker's intention to refer to a domestic house cat or a large jungle cat. The stop sign is not merely a command to halt the forward progress of a vehicle. It is also associated with a set of complex expected behaviors regarding what drivers are supposed to do when they arrive at an intersection marked with such a sign at more or less the same moment.

Other signs are considerably more difficult to discern for any number of reasons. First of all, the ideas or concepts manifest by the sign may be complex and difficult. Such is clearly the case with a sign like "entropy." A related but distinct problem can arise if the concept depends a great deal upon the context of its application. "Love," for example, is a sign that has different valid uses in many kinds of situations. It is also entirely possible that the cultural or scientific consensus that might support an unambiguous understanding and use of a sign is lacking. "Democracy" is a sign whose signification is often at the center of political conflict regarding whether a given polity can legitimately be described by the signifier *democracy*. The concept of democracy is one open to a wide range of potentially contesting interpretations. Another problem arises if we employ a sign in order to communicate about an effect or meaning that we experience, but we lack an explanation for this effect or meaning. In this case, there are attributes or elements to the signified that are unknown to us.

This brief excursion into the difficulties of interpreting signs does not exhaust the reasons why we have problems assigning meaning to our experience or communicating with one another, but it does point to the idea that the life of signs in society and culture might itself be complex, sometimes twisted, and relatively autonomous. Despite our best efforts to control them by imposing meaning on them, no one meaning, no one understanding, or no one positing of a signified may prevail. There may even be active competitors for the job, allowing the sign to take on a life of its own. Rarely, if ever, can Humpty Dumpty's famous assertion stand. He claimed that he could make a word mean anything he chooses as it was merely a question of whether he or the word was to be the master. Still, there was a certain final reality to his being 'broken' that allowed neither the King's horses or men to restore Humpty Dumpty to wholeness. The sign's autonomy can never be absolute, but the degree of its freedom is relative to the number of different signifieds associated with the same signifier, and how different they each are from one another.

"Information" is a sign that possesses a degree of relative autonomy from whatever language and culture sustains its use. The signifier *information* can be and is associated with a number of different signifieds, and some of them, although not entirely unrelated, are quite incompatible with one another. The ambiguity this condition creates makes it difficult to interpret the sign "information," and subsequently causes problems of mis-communication and misunderstanding. By extension, when we speak of "information science," we speak of a discourse that gives meaning to the sound image/word *information*.

Information science itself has already taken the first step toward this end by means of generally eliminating from its consideration phenomena to which the signifier *information* could refer that are not semantic in nature. Its next step is positive. It generally asserts that whatever is signified by the signifier *information* necessarily engages human communication. Technology plays an important role in information science, and it rightfully draws attention to itself. It offers a great promise for solving practical problems of access to information, but ultimately information science has declared itself to be a social science. There is a certain irony here. The sign "information," employed and explored by information science as a matter of semantics, clearly has something to do with human communication and so with signs and the relations between signifiers and signifieds. Thus, information science might very well be an important branch of what Saussure identified as semiology, a science of signs. Even with Fairthorne's assertions, there are still problems we must confront.

Contextualizing "information" by its role in human communication still leaves considerable ambiguity surrounding this central concept of information science. Knowing that users and uses are important theoretical objects for information science raises even more questions about what features and behaviors of these objects are important and how to observe them. As a matter of coping with the range of questions it sets out to answer, information science has historically displayed some general characteristics.[8] It is interdisciplinary in origin and nature, and its relations with other disciplines are in a state of constant change. The fields of study relevant to information science are themselves in a state of more or less constant change, and new fields of study are evolving with implications for information science. For example, the last decade of the twentieth century witnessed a rapid development of both cognitive science and semiotics, in turn promising fresh insights for information science. In addition, information science is clearly related to and driven by developments in information technology and computer science.

Despite the fact that information science borrows heavily from other disciplines and clearly manifests a technological imperative, the central intellectual purposes of the discipline involve the social nature of information as a vital human resource. The challenges associated with organizing and retrieving information, for example, are central to the solutions of problems that characterize human existence. Tefko Saracevic claims that information science has a unique role as an active and deliberate participant in the development and evolution of an information society, which necessarily means that it must be more than merely a technological discipline. He offers the following definition, and in the process, deploys a number of keywords (indicated by italics) that connote general problems for research and practice and signify the boundaries of the discipline:

> Information science is a field devoted to scientific inquiry and professional practice addressing the problems of *effective communication* of *knowledge* and *knowledge records* among humans in the *context* of social, institutional, and/or individual *uses* and *needs* for *information*. In addressing these problems of particular interest is taking as much advantage as possible of the modern *information technology*.[9]

Now compare this definition with that of Jesse Shera.

> The role of the library in the communication process and in the civilization that process serves is to *maximize the social utility of graphic records*. This is the standard against which all librarianship must be judged. The key words here are *utility* and *graphic records*, use and

> books. Thus librarianship is bibliographic, not bibliophilic. It seeks to unite in a fruitful relationship the book and the user. This does not exclude the possibility of sentiment or emotion in such a relationship, but it is the utility of the tool and the skill in its manipulation that justify an art, a profession, or a craft.[10]

Communication, and the relationship between a text and its user are central to both of these definitions. Both identify a number of very complex problems and deploy a number of words whose meanings are not free of ambiguity. Note that *text* like *information* can have many meanings. For our purposes, we will define *text* as a collection of signs purposefully structured by its producer with the intention of informing its consumer (leaving aside for the moment what it means to be "informed") whereby a collection of signs can appear in a variety of formats and media, including but not limited to writing. [11] "Text," in this sense, can substitute for both Shera's "graphic record" and Saracevic's "knowledge record,"thus rendering the difference between roles of information science and librarianship irrelevant. We are then left to determine whether these practices constitute a science, art, profession, or craft. There are reasons to believe that both information science and librarianship engage all four of these practices.

Thus, Shera dreamed of a new discipline he called *social epistemology*.[12] He envisioned this discipline to be a study of the "production, flow, integration, and consumption of all forms of communicated thought" throughout society in order to arrive at an understanding of the relation between knowledge and society so that society might make better use of what was collectively known.[13] This discipline would develop a theoretical body of knowledge and practical applications derived from that knowledge. Through understanding of how entire societies come to "know," one might transcend individual knowledge for the purpose of achieving social goals. Shera also insisted that this discipline be conducted on a genuinely empirical basis, grounded in the methods of social scientific research, and free from the bias of teleological definitions and assumed relationships.[14] Social epistemology, then, would focus on two central theoretical objects. One could study the situation of information use, asking questions about the context of information needs, the kind of information needed and available in given situations that involve human problems, and the nature of information use. Or, one could study the information unit itself, its nature, and its behavior.[15]

Shera's notion of social epistemology is not all that different from Saracevic's definition of information science. If the task of librarianship, or the information professions in general, is to identify, collect, organize, and make accessible, in logical and intelligible ways, the textual record of human experience for the purpose of advancing knowledge and improving the conditions of human life, then the task of information science is to systematically and empirically study these phenomena in order to develop a theoretical understanding of them, to improve the practices of the information professions, and to contribute to the achievement of improving the conditions of human life.[16] Thus, information science, although intimately engaged with technology, must ultimately be regarded as a human science. This condition, however, gives rise to a problem.

The Elusive Nature of Information

One of the most troubling questions facing the human sciences is whether or not the objects of their study are theoretically stable. A text, for example, is a relatively stable material object. Once the signs that constitute a text are assembled and given a physical form, the text is unchanging. A book will not rewrite itself. A painting will not repaint itself. A building will not rebuild itself. History will not literally repeat itself. With digital written texts, this condition is relative to the extent that they can be modified by their authors or by others. Printed books or paintings can be defaced, buildings can be modified or torn down, history can be re-interpreted, but in each of these cases one can argue that the original object has simply been replaced by a new object, which will retain its given form until modified again. The text, however, is merely a representation of meaning, of the content conveyed by signs, and the stability of meaning is considerably more doubtful. Just as Magritte suggested that his painting of a pipe was not, in fact, a pipe, a text cannot be regarded as the equivalent of the information that it communicates.

What is "information"? Is it the text or the content? The signifier or the signified? Does it even matter? A problem for information science is that its central theoretical object, information, is not entirely stable. This condition introduces ambiguity regarding what to observe and how.

Let us return to Saracevic. While his definition of information science incorporates concepts from a number of disciplines, each is a keyword in a manner reminiscent of Raymond Williams. Williams was not an information scientist, but he would have recognized these words. According to Williams, keywords might convey meanings assigned by philosophy or science, but they can also characterize and define the cultural life of a society in which they are used.[17] In turn, since their meanings are historically determined and cannot be easily and permanently fixed, these cultural meanings are often contested. The outcomes of these contests determine not only the meaning of the words but, to a great extent, the nature of the society in which they are deployed. The meanings of keywords can vary by time and place and by the nature of the society they define and which defines them. This is not to say that the meanings are arbitrary and without historical consistency, but rather that they are at the center of social conflict.

For example, our dictionaries customarily distinguish differences between wisdom, knowledge, information, and data, implying a hierarchical relation among them.[18] Data is processed to become information, which is communicated and manipulated as an object of cognition to become knowledge, which in its turn is contemplated and may become wisdom. This approach to these concepts, however, assumes that the criteria by which data, information, knowledge, and wisdom are defined are fixed, exhaustive, and unambiguous. But at what moment and how does information become knowledge? Can wisdom be derived from false knowledge, or is all that we call knowledge true by definition? The relationship between knowledge and truth is especially problematic. Information science is trying to solve some very complex problems, rendered more difficult by virtue of the fact that the words it uses to identify the theoretical objects of its investigations also identify a cultural and historical domain.

One promising approach to the conundrum is simply to cease worrying about "fixing" *information* so as to provide an intellectual focus for information science, and embrace instead the idea that at least there will be times when it can be unambiguously identified as being of interest to the discipline.[19] While there may be disagreements over its exact nature, information science assumes that information, like other natural phenomena, is a rationally intelligible object. Information, then, is an object whose nature and behavior can be empirically described, perhaps quantitatively measured, and certainly categorized and understood. Closely related to this view of information are phenomena associated with the means by which information can be represented, stored, and retrieved; and the goal of accurately and adequately representing what texts are about for the purpose of accomplishing these ends is of primary importance. The document, rather than other forms of text, tends to assume a privileged position in this view, and by offering the document as a primary object of interest to information science, it provides a logical and plausible way of defining the scope of information science.

Conversely, information is viewed by engineers and computer

scientists as an object manifesting certain characteristics which allow it to be controlled and manipulated by technologies to which its content and meaning, while preserved in the process, are nevertheless irrelevant. They aim to find and exploit the most general principles of information as a control phenomenon in order to design and build machines that will overcome nature's tendency toward chaos.[20] As information scientists, however, our success will depend on our ability to identify our needs, as well as our desires, and to understand the distinction between them. An information scientist must focus on the purposes for which we collect, organize, store, and retrieve information, and these purposes can, and often do, reveal a contradiction between needs and desires. In either of these cases, we must look beyond our information needs to the personal and social contexts from which they are derived. At this point, issues related to the use and users of information take center stage, and we must keep in mind that the study of information and its management is a means to the end of solving the general problems of human existence.

Different ways of thinking about information as a phenomenon are closely related to different ways of thinking about what needs to be studied and why. Here then is the approach we must take. By addressing phenomena of interest rather than attempting to define information, we can say that any understanding of information science must necessarily rely on a pluralism of theoretical approaches and methodologies. While some may regard this condition a strength, it may also be taken to mean that information science suffers from a "theoretical sterility" on one hand or on the other, a "flowering of the speculative imagination" whose fragmentation and imprecision lead it nowhere in particular.[21]

Indeed, can such an ambiguous phenomenon as information be the object of a scientific study? Is a science of information possible? To the extent that we agree that science is the means by which human beings explore and come to understand the ambiguities of nature, the answer is yes. The fact that information science must cope with an ambiguity associated with the concept "information" need not deter us from attempting to determine what information is, how it behaves, what it does, and what it means. Information science, then, broadly defined, is an intellectual discipline devoted to resolving information's ambiguities in order to make this critical human resource accessible and useable.

The Necessity of Multidisciplinarity

As we have seen, information demands both interdisciplinary study and multi-disciplinary solutions; not surprisingly, the parable of the blind men and the elephant comes to mind. In the story, each man examines a different part of the animal and inevitably arrives at a conclusion about the whole that differs from the conclusions of his colleagues. In the end, they still do not know what the elephant looks like, although their collective reconstruction of combined evidence yields a closer understanding than any of their individual efforts.

While this is a promising analogy for thinking about the nature of information science, it is an uneasy one. In order to assemble their particular observations into a coherent understanding, each man had to interpret the evidence before him, which in turn implies the possibility, if not the likelihood, of disagreement. In addition, as each is likely to insist that his particular view provides the foundation from which the overall interpretation should be constructed, there is bound to be tension. Finally, we really should consider the likelihood that since we are relying on the observations of blind men and parts of the elephant will be completely missed, some measure of the experience will be unacknowledged.

Let us now think of these men as information searchers. As Jesse Shera has noted, the multi-disciplinarity that characterizes the study of information began in the nineteenth century and is now reflected in a proliferation and fragmentation of groups devoted to that study, each organized around different intellectual principles and approaches to the problem.[22] Thus, information science evolved from documentation which evolved from librarianship, and even though information science is a relatively young discipline, the questions to which it seeks answers are not new.[23]

While information science certainly shares with librarianship the social role of securing the effective use of the human graphic record of late, more modern disciplines have become increasingly more important to information science.[24] The proliferation of information technology, the convergence of computing and telecommunications, and the rapid expansion of the information sector of the economy have all contributed to the fact that computer science is now a close affiliate of information science. Cognitive science—itself a cross between psychology, neurophysiology, cybernetics, and linguistics—contributes significantly to information science's search for a means of modeling human information seeking and processing and building artificially intelligent information retrieval agents. Finally many researchers in information science have recognized the close connection between information as a phenomenon and communication as a process and seek to integrate the work of the communications sciences into the discipline.[25]

The cross fertilization of information science with these other disciplines has strengthened its theoretical foundation, allowed old problems to be examined in new and useful ways, and identified new and important problems. But here too, tensions continue to abound. Are information retrieval and information use fundamentally technological issues or human issues? Are they issues of text or issues of content? And what is the nature of the relation between these two aspects?

Saracevic's argument is that the human/technology relation represents the primary unresolved philosophical, scientific, and professional issues of information science. He notes that the discipline is showing a tendency to devote greater attention to the human aspects of information retrieval, but wonders if too great a focus on technology is still causing the human dimensions of the problems of information science to go unaddressed. We are just now discovering how little we know about the human aspects of knowledge and information, he says, and perhaps we have reached a plateau in information retrieval research because we lack that understanding. Nevertheless, technological change draws attention to itself and away from the human aspects we need to study to reach the next level. His fear is that without clear research and social goals founded on a developed and robust philosophy, information science may be reduced to discipline that eases the adaptation of humans to machines rather than a discipline that finds ways to adapt machines to human capabilities and needs.[26]

Vannevar Bush's ideas, introduced to the general public in the late 1940s, acted as powerful stimulants to the scientific imagination, and they represent significant and important steps toward the development of information as a scientific concept. His notion of the "memex," an information storage and retrieval machine, based on the principle of associations between ideas, and available on a scientist's desktop, can be seen as a prediction of the Internet.[27] These ideas, however, also caused some problems. In particular he displayed a tendency to characterize human communication as a mechanical process whose mysteries rely on a logic which we have not yet discovered. His vision of information suggested that information retrieval systems be created that mimic the human brain. While such a metaphor possesses great suggestive power and, in fact, has stimulated creative theoretical research and innovative practical applications, the notion that human brains are merely processors of information is fraught with danger. It is a potentially short step from conceiving thought and communication as mechanical processes to embracing the idea that human existence is reducible to the self-regulating actions whose inefficacies are amenable to technological solutions. A technology may be judged effective on its own terms, for example, because it can accomplish what it is designed to accomplish. It may even be judged to be efficient in its work. But does it actually satisfy a human need or solve a human problem? We can invent technologies that will do a number of wondrous things. The question that technology cannot answer is, should we?

Systems of information retrieval are conventionally judged by their ability to adequately represent and provide access to information, but we can also ask questions about the quality, veracity, utility, and relevance of the information retrieved.[28] In short, by its ability to communicate information and knowledge in terms of the effect the system, or the information it delivers, has on its users, which criteria are most important, and does the answer to this question depend on who uses a system and why? At first, it may appear as if questions of evaluation and effectiveness are technical questions and that it may be possible to develop objective methods of answering these questions. Criteria of effectiveness, however, are not absolute, and this reflects a fundamental tension that influences our ability to determine a best way to evaluate systems of information retrieval.

At the heart of this issue is the relation between information systems—understood broadly to include everything from ordinary public libraries to commercial databases to the Internet—and the value conflicts that characterize the society in which these systems are constituted and operate. Information and its control can be used to achieve a variety of ends, and effectiveness, as a category of the evaluation of human activity, cannot transcend its indeterminacy because of questions and conflicts about the ends of human activity. An information system, for example, might be judged to be very effective at excluding users. In the context of a society that values the free flow of information as central to its identity as a democracy, such a system may be judged to be ineffective or even morally reprehensible. But what if we are told that the system is a high-bandwidth information/communication system for use by scientists and that, if opened to other kinds of traffic, its purpose would be impaired? Does everyone need the same level of access to every information system? If some potential users are excluded, how do we justify that exclusion?

These questions contain elements which require not only objective assessment but value judgments, and they point to yet another set of tensions identified by Saracevic. Information science is a discipline whose identity, concerns, and research agendas arise within the context of a complex information ecology consisting of:[29]

- Producers of knowledge

- Institutions
- Funders
- Publishers/Gatekeepers
- Dissemination channels
- Repackagers/database producers
- Libraries/information services
- Users

Each element represents a different set of stakeholders with different interests, views, goals, values, and ideologies, which, when taken together, constitute a setting where tensions and conflicts arise—a definite likelihood when "information" can mean something different to each of them. Technology both links and increases the value of each element to all, even as it increases the stakeholders mutual dependence and exacerbates tensions between them. According to Saracevic, solutions to information problems must engage the diversity of interest represented in this ecology, as well as respond to the contested social context within which the stakeholders live. Thus, the evolution of the information ecology can and will proceed without information science, regardless of its effectiveness (real or perceived).

The implicit warning in the above scenario is that the human aspects of these problems must take precedence over their technological counterparts if these answers regarding questions of effectiveness are to be found. The desire to focus on the latter is both strong and widespread. But the desire to reduce the complexities of human existence to mechanical simplicity is shortsighted. A "will to technology"is especially manifest in disciplines and practices that depend on technology. True, technology can solve many problems and, indeed, may be the only means of doing so, especially if the problem was itself a creation of technology. The problems of effectively organizing, finding, and retrieving Internet-based information, for example, will undoubtedly require technological solutions. This condition, however, carries with it a great temptation to view the human aspects of information problems through the lens of technology or worse, overlook them altogether. The solution of these problems will require a multi-disciplinary approach, despite the theoretical confusion that is likely to result.

Each discipline of inquiry and practice that can be identified as a component of information science—librarianship, cognitive science, computer science, and communications science—can and often does offer different meanings of the keywords in Saracevic's definition of information science, and most of these meanings differ yet again from common, everyday understandings of these words. This is especially true for the word *information.* As with the blind men and the elephant, each observer can offer a plausible interpretation of the elephant's nature but none can grasp the whole. It is undoubtedly true that information science must necessarily be multi-disciplinary because the object of its study, information, is a complex and multi-faceted object. The various views of information represented by this disciplinary fragmentation, however, tend to support Fairthorne's observation that the object itself appears to be so unstable and ambiguous that we must sometimes entertain the possibility that the blind men are not examining different parts of the same elephant, but rather have hold of a multiplicity of animals.

Problems of Metaphor and Perspective

Given the ambiguities and tensions that characterize the phenomena that information science seeks to investigate, what kind of science can information science be? It evidently must be a science of both material, informative objects and the interpretation of their meaning, of things and of people. But is there not an irresolute tension between the way we study and know things and the way we study and know people? Do we treat information as a thing, created by any number of processes, including human thought, and grant it an independence of behavior whose regularities are like those of any other natural object? Or as a uniquely contingent and transitory phenomenon whose existence depends upon a moment of deliberately communicative interaction between two or more human beings? To slightly twist an old aphorism, if information falls in a forest and no one hears it, does it make a sound? Does it inform? Even if someone does hear it, but it makes no difference, is it information or noise? If one person hears it and another doesn't, whose perception do we accept concerning its status as information? These questions imply that the nature of information can be characterized by a duality that simultaneously manifests determinate and indeterminate qualities, and this condition poses problems for how we might come to know and understand its nature.

Science presupposes a reality separate and apart from what we perceive as real. In this view, reality goes on being what it is, doing what it does regardless of what we know or don't know about it, or even whether we are aware of it. The occurrence of an event can be accepted as a scientific fact if more than one person can observe and describe it in the language of physical things: "colors, sounds, pointer readings, weights, and so on."[30] However, this description must also be reliable, in that many different observers should describe the event in much the same way. To meet these criteria, science employs operational definitions of the phenomena it seeks to observe by describing how an observer is to acquire his or her facts about them.[31] Thus, this kind of reality is immune to our values and our biases which may distort our judgments of reality. Instead, reality is constituted by real objects whose existence and behavior are independent of our perceptions of them.

Based on our observations of such objects and their behavior, we can develop theories to explain them and attempt to manipulate them to achieve predictable outcomes. If our theories are confirmed, we can claim, at least tentatively, an objective knowledge of reality. The more often our manipulation succeeds, the more confidence we can have in our knowledge, on the assumption that "since an event has occurred before on several occasions, under similar circumstances it will happen again."[32] If our manipulations fail, we learn what is not likely true of reality. Sometimes our manipulations are successful, but we later discover that the reasons why this is so are not all what we thought them to be. Sometimes we can apply our predictions regarding the nature and behavior of real objects to the development of technologies that have an effect we intend, even though we really don't understand fully *why* that effect occurs.

All of the above happens because the methods we use to manipulate objective reality are grounded by a language of physical things that posits and categorizes the objects constituting this reality as separate and apart from our perceptions of it. Regular and constant in their nature and behavior, they dutifully obey universal, natural laws of cause and effect, whether or not we fully understand their purpose.

Many of the elements of the information ecology described by Saracevic, for example, demonstrate quite consistent observable behaviors. In terms of information, documents and records of documents are the traditional, print-based containers of information, the objects that circulate within and sustain the information ecology. They are the physical matter that is produced and exchanged by the elements of the information ecology. Even as technology makes it possible to convert the content of these containers to electronic form and so introduces possibilities of storage, retrieval, transfer, and exchange that until recently could not have been imagined, the content still retains a physical, albeit not entirely tangible, form.

Given the material reality of documents, as well as other texts, and our ability to manipulate them by means of electronic technology, there is every reason to believe that science offers an appropriate means to study phenomena such as the elements of Saracevic's information ecology. Institutions, publishers, dissemination channels, libraries, information services, and information users all possess a historical character and manifest aspects which can be observed and measured. In a way, progress in any science presupposes a faith based on our ability and past success to discover more and more about material reality, to refine our theories, and to develop ever more accurate representations of reality. Information science stakes its truth as a science on these same grounds. Its claims about the nature of the reality of information inform a technology that makes possible the development of working mechanisms of information storage and retrieval.

But can we say that information is unambiguously separate and apart from our perception and experience of it? What if, instead, it is a phenomenon whose reality vitally depends on our perception and experience of it? What if information is an aspect of human reality that is not an external given, but is instead psychologically and culturally constructed within and by human consciousness, and cannot be grounded entirely in a language of physical things? Consider for a moment two crucial elements of the information ecology: producers and users of knowledge. Can the diversity and variety of human beings, and the motivations, intentions, desires, and behaviors these terms represent be so reduced as to become merely elements in a system? Are they so easily knowable and predictable?

It is true that we have successfully applied scientific inquiry to study, predict, manipulate, and change the phenomenon we call information. We have developed technologies that allow both producers and users of knowledge to become more effective and efficient. But still we are working within a metaphor; and our need to employ a metaphor such as "information ecology" also reveals how little we really know. The "information ecology"could just as easily be referred to as an environment, a system, a mode of production, or a site of human struggle. Indeed, each of these metaphors has been employed to guide useful scientific inquiry about the nature of information and information problems, yet none comes close to the elephant of truth discussed earlier. Since definitive knowledge about human affairs eludes us, we instead tend to invent more or less useful metaphors that not only represent human reality but, in the process of representation, create that reality.

Unlike our knowledge of material reality, our knowledge of human reality seems to change rather than accumulate. Even now, despite decades of study we still lack fundamental laws of human science that compare with the certainty of the laws of physics. Perhaps this is because the objects that constitute human reality may not be entirely stable or entirely describable in a language of physical things. Consequently, the social sciences often reify the objects of their study, that is they treat a phenomenon that is not exactly a thing as if it were one.[33] This strategy tends to reduce complex human realities to their simpler aspects in order to observe them in the controlled fashion that the objectivity of science demands. While often successful, it is a process that often impoverishes the reality under investigation.

Libraries, for example, are often conceptualized as document delivery systems in order to study how well they serve to expose and connect documents to users.[34] The study of these objects, including library users as objects that seek and find documents, has usefully contributed to an understanding of how libraries work and why. This knowledge, in turn, has been employed by library managers to improve the effectiveness and efficiency of library services. Still, there are aspects of the library as a human institution that are not and cannot be revealed by this kind of investigation; for a library is not merely a system of interacting objects. It is also constituted by the daily interaction between library staff members, between the staff and the library's users, and between all of these people and the objects from which access to information arises. In addition, all of the above occurs in a historical and cultural context that provides meaning to their existence and interactions and conditions the relationship of a library to its particular community.

From this point of view, a case can be made that when the library closes at night and the last person leaves, it ceases to exist until the people come back and it reopens in the morning. In the case of digital libraries, we might concede that the library exists as long as the servers are turned on, but we might have to concede that only the action of a human being logging on to those servers actually calls it into existence. Any library possesses certain objective characteristics that can be unambiguously observed, but it is also a social and cultural construction whose reality is made and re-made from moment to moment by human beings whose desires, values, motives, actions, and interactions endow it with meaning. From this point of view, the effectiveness of a library has less to do with its ability to connect documents to users than with how and why it serves to produce and reproduce a particular kind of culture.

By introducing the idea of meaning and its relationship to the construction of human reality, we finally begin to approach the issue at stake. Ordinarily, an object or phenomenon is defined by the way its measure is taken as a physical thing. The language of physical things is an operational language, but issues of meaning in human affairs are considerably more difficult to resolve, and indeed, these affairs tend to proceed even if their meaning is unestablished, unclear, or contested. In fact, human affairs themselves are often about contests of meaning.[35]

Now let us recall Saracevic's conception of information science and Williams's notion that keywords are words for which final and definitive meanings are elusive. Information science may deploy these words with reference to certain understandings of the realities they represent, but it will face contenders who will also assert a claim on these words and their meanings. Information science involves human agents, each of which can, and often will, assert a different meaning for each of Saracevic's keywords. Other agents will be in other academic disciplines. Some will be other organizations, institutions, and perhaps political interests. But the terms each uses to identify itself and each other will also be indicative of a power struggle of some sort. Information science may claim that its identification of information phenomena rests on a scientific and objective basis, but this will make little difference to agents that do not recognize its legitimacy.

Conflicting claims to meaning are to a great extent the source of the tensions that Saracevic says are inherent in the information ecology. We have to realize that what we are dealing with is less an ecology and more a social field on which one small part of human destiny is contested and controlled. Unlike a real ecology, a metaphorical one is created from human intention and social construction. The information problems that characterize this ecology may not be so easily resolved if their articulation is based on desire disguised as objectivity. The result, as we learned from Humpty Dumpty, is that a word may mean anything we want it to mean, but no words spoken can repair a shattered egg.

Problems of Information Science

The problems that information science seeks to solve manifest diverse features and suggest a number of plausible approaches. As we have seen, the discipline has developed a multi-disciplinary character, but this condition may be due to more than the complexity of the problems information science seeks to solve. There seems to be something indeterminate about the object of its study that imposes a fragmentation of thought. This indeterminacy gives rise to substantial disagreements about which disciplines, and which meanings of "information," should be given precedence as being central to information science, and while there is widespread agreement on the idea that the purpose of the discipline is to provide solutions to human problems, disagreements abound concerning the nature of its fundamental theoretical object.

The indeterminate quality of "information" as a theoretical object derives from conditions that impose on the word a need to convey different meanings in different contexts. Ultimately, the different contexts are sustained by differently experienced realities. In information science for example, there are at least two theoretical realities: one in which information is regarded as a material object, and another in which it is regarded as a cognitive phenomenon. One "information" is "out there," independent of human consciousness, and the other is "in here" as a creation of human consciousness. Again, it is vital to stress that the difference here is more a matter of experience, discourse, and relative emphasis than it is an absolute distinction.

These theoretical meanings also compete with ordinary cultural and historical meanings. Producers and consumers of information, for example, may substantially share the same reality, which is why particular consumers seek out information created by particular producers. In both cases "information" may be a means to an end, but their differing ends are mediated by and dependent on their unique status as producer or consumer—in other words, on the desire to inform or become informed. Even when producer and consumer share the same topical interest, each may each be motivated by different problems. All of the meanings of "information" presented so far may be very different from how it is defined for a management information systems analyst and a publisher, to select just two possible members of an information ecology. For the former—whose concern is to get a package of information from producer to consumer in the most efficient manner—or for the latter—whose concern is with information as a commodity whose sale sustains a business—content, while not entirely irrelevant, is not central to their meaning of "information." For the producer and final consumer of information, however, content is the primary concern.

From all of these examples it is clear that "information" is not indeterminate because its nature and value cannot be defined, pinned down, or described—not because its reality cannot be determined, but because its reality can plausibly and usefully be determined in so many contradictory ways. We do not yet understand how these different determinations fit together so as to generate universal solutions to the problems of information storage, retrieval, and use. One of the few things we can say about "information,"at least in its textual manifestation, is that it is a residual creation of human existence—a record of what we do, what we've done, what we think, what we know, and what we think we know. This residual creation, in all of its forms, contexts, and uses is the focus of information science's attention.

As a science, its broad goal is to generate knowledge of information—to understand what it is, what it does, how it behaves, and why. Additionally, it seeks understanding of the use and users of information. Like any science, the practices that constitute information science may not yield an immediate utility, but in the end its purpose is to create and enhance a means of controlling information in order to rationalize the relations between information and human affairs. An ancillary but very important purpose is to contribute to conditions of rationality in human affairs by rationalizing the use of information in of human affairs.

Historically this problem has been known by many names, including "bibliographic control" and "documentation," but both of these signs are culturally coded in a way that associates them with particular practices and institutions at particular times and places. A more general term that embodies these practices, times and places, as well as other is "access." To bring together the goal and premise of information science, then, we can say that it is about generating knowledge that will enhance access to information. "Access," however, is itself a complex sign, made more so by its dependency on the indeterminacy of "information."

"Access" implies, and as a practical matter embraces, our knowledge of information as a theoretical object. Access is a practice based on that knowledge. At the very least, "access" signifies retrieval—finding and getting information. In other words, it describes the means of controlling information for particular human purposes, but control is more than a matter of merely finding and getting. It is possible to retrieve the wrong information, information that is inaccurate, inadequate, inappropriate, misleading, or even damaging and oppressive. It is possible to retrieve information that diverts, dissimulates, and distracts. It is entirely possible, through the use of logical and effective means of access, to retrieve information that inhibits rational choice and contributes to nonrational or even irrational choice. If access is to be successful, it must be a matter of finding and getting the right information, and there are two critical aspects to the solution of this problem: representation and relevance.[36]

"Representation" signifies the theoretical problem and practice of organizing information and making it accessible for retrieval. The amount and diversity of the information record precludes easy access. Even in a small library, it is very difficult to retrieve a particular book unless all of the books are arranged on shelves in some logical and knowable order. In home libraries, many people organize their books in alphabetical order by the author's last name. In libraries, it is more typical to arrange the books first by class number that ostensibly groups together books about the same topic, and then within a given class in alphabetical order by the author's last name. However, organization alone is not enough, especially as the amount and diversity of information to be stored and retrieved increases. In a small library, browsing can be a very effective way of retrieving information, but it becomes quite impractical as a collection size increases, to say nothing of the problems arising from the possibility that the needed information may not be physically present to retrieve. The solution is an intermediary. Library catalogs and journal indexes, whether print or electronic, are examples of such devices. They are files of text surrogates that represent the texts in a collection. Representation of this kind engages two fundamental aspects of the texts represented, one of which is significantly more important to easing the problems of access, although both are necessary.

Descriptive representation is the means by which a text surrogate is created for inclusion in a file. This action is a matter of establishing the existence and identity of a particular text in such a way that the surrogate represents only that text and no other. Perhaps the most familiar example is bibliographic description of books for the purpose of creating library catalog records. These records typically include, among other facts regarding a book or other text, the book's author, title, publisher, edition, and place and date of publication.[37] Bibliographic control—indeed, information control of any kind—begins with the accurate and adequate distinction of the identity of one text from that of any other. To create access to the right information, issues of identity must be sorted out, but this condition also reveals a most intriguing irony.

We conventionally assume that successful access is a matter of retrieving information that best matches our need for information. The issues of descriptive representation, however, suggest that what does not match may be at least as important, if not more important, to successful retrieval. Topical representation is the means by which the subject of a text is identified. Through the use of such devices as classification, subject headings, descriptors, and index terms, texts sharing a similarity of subject are grouped together according to the concept known as "aboutness." This form of representation is crucial to organizing information, as it is the means by which we create systems of organization and attempt to match the content of information with the source of the need for information, and it is a central concern of information science.[38]

More often than not, however, a seeker of information wants to retrieve particular information related to a problem to be solved, rather than simply retrieve a text on a topic. Certainly the former problem is the more difficult of the two, made even more so by the ambiguity of "information," and irony of difference begins to make its presence felt. A central concern of information science is to empirically investigate the general nature of the problem of representation, especially the problem of representing the aboutness of texts, and to compare the relative merits of various means of representation. But what, exactly, should be represented? The text or its content? Can we assume these phenomena are identical? If not, can the latter be assumed to be as stable as the former, and even if so, can the *meaning* of the content be granted this same status? Given the ambiguous and unstable nature of information, it should come as no surprise that "representation" suffers the same condition. At stake is not merely retrieving information about a user's need, but retrieving information that will make a difference to the user's life.

Adding to the confusion are three potentially competing points of view from which the aboutness of texts can be assessed.[39] The first is the author's view. Presumably an author creates a text with the intention of conveying a particular content and imparting a particular meaning. Second is the professional indexer's view who—with considerable knowledge of the subject of the text, other similar texts, and the various available systems of organizing texts— may exercise an independent and quite different judgment regarding the aboutness of text based on contextual aspects not considered by its author. Third, the user who actually retrieves a text may read into it a meaning based on unique and personal circumstances completely unrelated to those of the author or fail to retrieve a text that might be useful because the indexer's representation of its aboutness failed to match the user's understanding of his or her personal problem that motivated the search for information in the first place. Which view should be privileged for the purpose of access? Whose interpretation of the text is correct, and what criteria should be used to establish correctness? Once again, the problem of information, manifest in a question of whether its nature is bound up in its material and objective form or its subjective interpretation, arises, and this implies consequences for developing and actually advancing a stable body of theoretical knowledge about representing information.

Some system of organization, based on representation, is obviously necessary if one is to retrieve information, but once that step is taken, a second is implied. Not all information retrieved, even if similar, will be equally useful or related to the problem that prompted its search. In the moment at which all the knowledge of information science, including the practices informed by that knowledge, brings information to a user, that user must determine the differences between each bit retrieved and how different they all are from the need that motived the search for information in order to determine their relative value. At the very moment similarity does its work, difference comes into play. The implications are twofold: successful representation is as much matter of separating texts by their differences as it grouping them by their similarities, and in both representation and use, the second critical aspect of access, relevance, is at play.

Van Rijsbergen defines "relevance" as "the measure or degree of correspondence or utility existing between a text or document and a query or information requirement as determined by a person."[40] He goes on to acknowledge, however, that this measure is "inherently uncertain" and "context dependent." First, the representation of a text must be "relevant" to its content and aboutness, a determination is itself inherently uncertain. Second, a user of a retrieved text must determine it is useful, i.e., satisfies the need that prompted its search. A retrieved text may be topically relevant to a query in the sense that both text and query are about the same subject, but this condition does not guarantee that the use of that text will actually contribute to the solution of an information seeker's problem.

"Relevance," then, clearly signifies something about the user's unique and personal circumstances and the context within which he or she makes a determination of relevance. At the same time, a correspondence between text and query may have little to do with the correspondence between content and need. Here again, difference is at play. Two texts, each represented in the same way about the same topic, do not necessarily possess the same relevance. The key is the user's distinction of the differences between the two texts regarding content and meaning. A text that is likely about a given topic still may not contain the information needed. If "representation" presents itself as an ambiguous sign, then "relevance" raises even more problems, and if "access" as a sign depends on these two ideas, then its meaning and the stability of that meaning can be called into question.

Conclusion

Thus we come face to face with the dilemma of information science. How do we generate knowledge about information and access that will enhance access to information, where both "information" and "access" are inherently ambiguous and unstable theoretical objects? Remember, the indeterminacy of these signs and the phenomena they represent do not derive from the fact that they cannot be determined, but they can plausibly and usefully be determined in a variety of ways. Their interpretation and their meaning, indeed their reality, are contingent and context dependent phenomena. The question of which should be privileged can only be answered by the rejoinder, it depends on what you want to do.

David Ellis provides a useful tool to address the essential indeterminacy of information. I will explore his analysis of the nature and role of what he identifies as the physical and cognitive paradigms in information retrieval research, and his assertions regarding the existence of a "categorical duality" that characterizes the nature of information in the next chapter[41] I intend to use this tool in a modified form to explore the ambiguities of information. There are problems with the notion of paradigm in general, and they are acute when the notion is applied to the social sciences. One example of these problems is already apparent in "information ecology" as a sign applied to the people, things, institutions, interests, and needs that constitute the domain of phenomena of interest to information science. This sign derives its power from its suggestiveness as a metaphor rather than its status as a paradigm.

Looking at information as theoretical object through the lens of what I will call the physical and cognitive metaphors is more revealing than trying to apply the more restrictive notion of paradigm. The next chapter will explore this idea. The following four chapters will use the idea developed in chapter two to explore the ambiguous duality of information and provide examples of research based on each metaphor. Chapters seven through nine will explore the ambiguous nature of access through a discussion of representation and relevance. In each, the unity of information, necessarily deconstructed in order to cope with its ambiguities, will be restored. Both representation and relevance engage text as material object and content as an outcome of subjective interpretation. This condition leads to yet still more and different research questions and practical consequences, and reveals the complexity of the problems information science seeks to solve.

The picture will not be complete, however, without looking at information use and users. The consideration of relevance provides the means of transition to this subject in chapter nine. In that chapter and the next, I will explore the ambiguities and instabilities of use and users as theoretical objects, as well as how these particular ambiguities suggest an entirely different way of thinking about information that the original approach through the idea of a categorical duality does not allow. Finally, from the implications of this effort, I plan to take a critical look at an "information" that information science explores to a limited extent, but for some reason fails to fully engage. Another metaphor awaiting further exploration is information as a social phenomenon. This metaphor constructs text, content, need, and use as cultural and collective phenomena rather than as individual phenomena. In the end, this book is about exploring implications. We cannot really understand "information" as a theoretical object or solve the problem of access to information without engaging its multiple realities as text, content, and meaning; as simultaneously a medium and end of communication; and as individual and social phenomena. We are still at a point in our journey where it is not yet possible to fully engage one reality without ignoring, at least to an extent, the others.

For those of you who intend to become information scientists, the goal of this book is to introduce you to the fundamental concepts of your future work and to suggest reasons for the persistent conceptual problems you are likely to confront. Others may intend to become librarians, and for you I present what I think is a compelling reason you should know about information science. While librarians work with technology and use science to evaluate their work and understand users and their needs, that work is fundamentally about interpreting human situations and the role that information plays in those situations. Librarianship manifests many of the characteristics of an art, and so another goal of this book is to introduce you to the theory that underlies and supports that art. Yes, art can be practiced and practiced well without this knowledge. There are painters and poets, writers and musicians who never studied the science of their art, but a consciousness of the principles of line, the structure of music, or the grammar of language can and will inform the artist's work. Information science may not yet have all the answers it seeks, but it nevertheless represents our best effort to understand the problem of information.

Endnotes

1. Zhang Yuexiao, "Definitions and Sciences of Information," *Information Processing & Management* 24, no.4 (1998): 488-489.
2. Michael Buckland, "Information as Thing," *Journal of the American Society for Information Science* 42 (June 1991): 356. Italics in original.

3. Robert Fairthorne, "Information: One Label, Several Bottles," in *Perspectives in Information Science* ed. Anthony Debons and W. J. Cameron (Leyden: Noordhoff, 1975), 65.
4. Yuexiao, "Definitions and Sciences," 481.
5. Nicholas J. Belkin and Stephen E. Robertson, "Information Science and the Phenomenon of Information," *Journal of the American Society for Information Science* 27 (July/August 1976): 198. Italics in original.
6. Ferdinand De Saussure, *Course in General Linguistics*, ed. Charles Bally and Albert Sechehaye, trans. Wade Baskin (New York: Philosophical Library, 1959), 65, 66.
7. Jere Paul Surber, *Culture and Critique: An Introduction to the Critical Discourses of Cultural Studies* (Boulder, Col.: Westview Press, 1998), 159-160.
8. Tefko Saracevic, "Information Science: Origin, Evolution, and Relations," in *Conceptions of Library and Information Science: Historical, Empirical, and Theoretical Perspectives*, ed. Perth Vakkari and Blaise Cronin (London: Taylor Graham, 1992), 6.
9. Ibid., 11.
10. Jesse Shera, *The Foundations of Education for Librarianship* (New York: Becker and Hayes, Inc.; John Wiley & Sons, 1972), 194.
11. Belkin and Robertson, "Information Science," 20.
12. Jesse Shera, "Social Epistemology, General Semantics, and Librarianship," in *Libraries and the Organization of Knowledge*, edited and with an introduction by D. J. Foskett (Hamden, Conn.: Anchon Books, 1965), 12-17.
13. Ibid., 16.
14. Jesse Shera, "Foundations of a Theory of Bibliography," in *Libraries and the Organization of Knowledge*, 23-28.
15. Ibid., 30-32.
16. Shera, *Foundations of Education*, 164-179, 190-194, 197.
17. Raymond Williams, *Keywords, A Vocabulary of Culture and Society*, rev. ed. (New York: Oxford University Press, 1983), 15-17.
18. Brian C. O'Connor, *Explorations in Indexing and Abstracting: Pointing, Virtue, and Power* (Englewood, Colo.: Libraries Unlimited, 1996), 7-8.
19. Gernot Wersig and Ulrich Neveling, "The Phenomena of Interest to Information Science," *The Information Scientist* 9 (December 1975): 127-140.
20. James R. Beniger, *The Control Revolution, Technological and Economic Origin of the Information Society* (Cambridge, Mass.: Harvard University Press, 1986), 170-184.
21. Bernd Frohmann, "The Power of Images: A Discourse Analysis of the Cognitive Viewpoint," *Journal of Documentation* 48 (December 1992): 365-366.
22. Jesse Shera, "Of Librarianship, Documentation, and Information Science," in *Knowing Books and Men; Knowing Computers Too* (Littleton, Colo.: Libraries Unlimited, 1973), 272.
23. Ibid., 269-278.
24. Saracevic, "Information Science," 13-18.

25. William Paisley, "Information Science as a Multidiscipline," in *Information Science: The Interdisciplinary Context,* ed. J. Michael Pemberton and Ann E. Prentice (New York: Neal-Schuman, 1990), 3-24.
26. Saracevic, "Information Science," 21.
27. Vannevar Bush, "As We May Think," *Atlantic Monthly* (July 1945): 101-108.
28. Saracevic, "Information Science," 21-22.
29. Ibid., 22-24.
30. Garvin McCain and Erwin M. Segal, *The Game of Science*, 2nd ed. (Monterey, Calif.: Brooks/Cole Publishing Company, 1973), 45.
31. Ibid., 46-47.
32. Ibid., 67.
33. Peter L. Berger and Thomas Luckmann, *The Social Construction of Reality* (Garden City, N.Y.: Doubleday Anchor Books, 1967), 89-92.
34. F. W. Lancaster, *The Measurement and Evaluation of Library Services* (Washington, D.C.: Information Resources Press, 1977), 1-17.
35. Irving Goffman, *The Presentation of the Self in Everyday Life* (Garden City, N.Y.: Anchor Books, 1959), 1-16.
36. Peter Ingwersen, *Information Retrieval Interaction* (London: Taylor Graham, 1992), 49-56.
37. Ibid., 53.
38. Ibid.
39. Ingwersen, *Information Retrieval*, 50-52.
40. C.J. van Rijsbergen, "The Science of Information Retrieval: Its Methodology and Logic," in *Conference Informatienvetenschap in Nederland* (Haag: Rabin, 1990), 24.
41. David Ellis, "The Physical and Cognitive Paradigms in Information Retrieval Research," *Journal of Documentation* 48 (March 1992): 45-64.

2

Paradigms and Metaphors

In the previous chapter we saw information science as a discipline divided within itself due to complexity of the problems it seeks to solve and the inherent instability and ambiguity of information as a theoretical object. There are good reasons to assume that information, and phenomena associated with it, can be conceived as natural objects whose reality can be described in the language of physical things. This idea directs theoretical attention to the objective characteristics manifest in and by particular material forms of information in order to arrive at general conclusions regarding the nature and behavior of information. On the other hand, there are equally good reasons to assume that these phenomena can be conceived as social and psychological constructions whose reality is situationally contingent. This idea of information directs theoretical attention to the subjective and cognitive effects of information as it informs in order to arrive at conclusions regarding the nature and meaning of becoming informed.

This duality is not unlike the one confronted by physics regarding the nature of light. Under some circumstances, in order to explain its behavior, light must conceived as a waveform. Under other circumstances, to accomplish the same end, light must be thought of as a particle. According to David Ellis, this ambiguity resolves into two broad and apparently conflicting approaches to the study of information that he identifies as the physical paradigm on one hand and the cognitive paradigm on the other.

In language very similar to that of Saracevic and Shera, Ellis describes information retrieval research as a discipline that:

> has as its research focus a system where people and artefacts are involved in a number of complex interactions—authors, indexers, abstractors creating texts, indexes, abstracts and databases which are then approached by a variety of different users, in different ways for different purposes.
>
> Underlying this complexity is a basic irreducible duality—people (authors, indexers, intermediaries, and users) and things or artefacts (documents, document representations, abstracts, indexes, databases). The physical paradigm takes as its primary focus the artefacts, whereas the primary focus of the cognitive paradigm is the people.[1]

Note that many of the concepts, relationships, and research problems identified by Saracevic and Shera in their respective descriptions of information science and librarianship appear again in this statement. Ellis, however, is more explicit regarding the essential ambiguity of information as a theoretical object, assigning to its nature an "irreducible duality," and linking this duality to an evidently necessary, if unavoidable, "categorical duality" of research focus in information retrieval. The unfortunate result, says Ellis, is "fragmentary, or piecemeal" research, which in turn inhibits the progress of information science as a "theoretical science within a well developed paradigmatic framework."[2]

Ellis's analysis of the existential condition of information retrieval research echoes Thomas Kuhn's groundbreaking and controversial work, *The Structure of Scientific Revolutions.*[3] The notion of a paradigm is central to Kuhn's explanation of how science works, and to how and why scientific thought and practices changes over time. Kuhn disputes the idea that science is a matter of the progressive and linear accumulation of knowledge, proceeding from essential and fundamental principles to more complex understandings and explanations of observable natural phenomena. Rather, our knowledge changes, if not advances, by means of revolutionary shifts in perspective and thought, signified by the emergence of new paradigms that embody new ontological assumptions about the nature of observed phenomena.

The classic example of such a shift is that from a view of the nature of the solar system that put the earth at the center of it to one that replaced the earth with the sun. The first view served well for quite some time, successfully guiding early astronomers in their explanation and prediction of the movement of heavenly bodies. Eventually, however, an accumulation of observations and unfulfilled predictions , which Kuhn calls anomalies, led to the Copernican "revolution" that put the sun at the center of the solar system. The new paradigm not only allowed both old observations to stand by resolving the anomalies but also ensured the permanent passing of the old paradigm.[4] In introducing the idea of "paradigm" to describe the nature of scientific practice, Kuhn suggested that a paradigm will dominate a scientific practice for a long period of time, that paradigm shifts do not occur quickly, and that once defeated by a challenger, an old paradigm will recede into the history of science, leaving its field to a new one.[5] He did not mean that every short term shift in focus or theory constituted a paradigm shift, nor did he believe that the existence of multiple or contesting paradigms could constitute a status quo or sustain "normal science."[6]

This conclusion raises some difficult questions with respect to Ellis's identification of two active and competing paradigms in information retrieval research, and calls into question the status of information science as a science. Kuhn writes,

> Throughout the pre-paradigm period when there is a multiplicity of competing schools, evidence of progress, except within the schools, is very hard to find. This is the period described in Section II as one during which individuals practice science, but in which the results of their enterprise do not add up to science as we know it.[7]

Notice that Kuhn does not say that a pre-paradigmatic, or "immature," discipline cannot or does not apply scientific methods of inquiry to particular problems. His implication is that such a discipline is not yet a "science." But even as Ellis concedes the latter, he argues that an immature science can have paradigms, and that an analysis of their role can be a useful means of understanding its structure and development.[8]

Ellis makes a reasonable case that information retrieval research displays a congruence with Kuhn's basic description of science as an intellectual and social practice. His identification of the different theoretical objects that occupy the attention of what he calls the physical and cognitive paradigms, as well as their implications for information as a theoretical object, resonates with much of what we observed in the previous chapter. Nevertheless, when Ellis concedes that the cognitive paradigm lacks a foundational experiment, or exemplar, it seems as if "paradigm" means one thing for Kuhn, and another for Ellis. According to Kuhn, an exemplar is a necessary starting point for a paradigm.[9] There is no doubting the ambiguity of the concept of "paradigm," but it remains powerfully suggestive. It can be a useful tool for understanding both the sources and nature of theoretical and methodological disputes within a discipline of inquiry, and it can help us to see just how discipline of inquiry exercises its discipline, but it must be used with caution.

Paradigms

In essence, a paradigm is an organizing principle, and a brief examination of how paradigms work to organize language can be helpful in understanding the application of this concept to the practice of science. For example, linguistics uses the word "paradigm" to refer to a pattern of inflection among nouns and verbs and to typical samples of these words that identify a system of inflection. A sample inflection, then, serves to exemplify the entire system.[10] Following this logic, it is not difficult to surmise that in science each experiment serves as an example of similar but not identical experiments, all addressing the same object of study in pretty much the same way. The key point here, is that a paradigm signifies a set of systematic relations between certain elements or phenomena. In the case of language, the elements in question are words.

Linguistics also employs "paradigm" in a Sausseurian sense. Language, he says, relies on two dichotomous sets of systematic relationships.[11] Syntagmatic relationships are a matter of elements present in a given linguistic unit. These are about the linear quality of language and the rules that govern the order of elements in linguistic units. For example, syntagmatic relationships exist between particular words in a sentence.[12] Paradigmatic relationships are a matter of elements in absentia. These relations signify a class of elements, associated by given similarities, that can occupy the same structural position in a linguistic unit, depending on the context. Paradigmatically related words, for example, can be substituted for one another in a given sentence, and the sentence will still be meaningful.[13]

Take for example the sentence, "I go out." These three words are syntagmatically related to one another and thus form a meaningful linguistic unit. The paradigmatic relationship between words *I*, *he*, *she*, *we*, and *they*, however, allow them to substituted for each other *I* in this sentence while still preserving it as a meaningful sentence. The potential substitutes are not present in the sentence, but because of similarities they share with *I*, they could be present and are potentially implied.[14] The logic of syntagm and paradigm is very powerful. It allows us to speak and understand sentences that we have never before spoken or heard, and it can be extended beyond linguistics to any circumstance that implies "choices made among classes of elements and functional combinations of chosen elements."[15] The sensibility of that circumstance will depend on how the elements of the situation are selected and combined.

Any process representing a syntagmatic combination of elements that moves horizontally from a beginning to an end, necessarily requires a system paradigmatic selection cutting vertically through the process in order to for the process to make sense.[16] The sensibility and meaning of a discourse, including that of science, depends on syntagmatic and paradigmatic relations among and between the elements that constitute the discourse. For example, in Kuhn's case, he acknowledged the syntagmatic in the practice of what he called "normal science," but his primary concern was with how science changes. He argued that through paradigm change, new elements, experiments, observations, models, and theories could be introduced into scientific discourse and old elements combined in new ways. The overall effect is to allow a scientific discourse to make new statements about the nature of reality.

Kuhn argues that paradigms precede theory; in so doing they serve to organize scientific activity by providing a way of seeing certain problems of research as possessing a similar nature that, in turn, may yield to similar solutions. Prior to the emergence of a paradigm these problems appear as different and unrelated, and consequently work on them remains isolated and uncoordinated. This is why scientific methods can be practiced in the context of work on the individual problems, yet the work as a whole not be considered science. To come into its own, a paradigm requires a specific scientific achievement to serve as its sign and provide the foundation for further scientific practice. As a concrete artifact representing the paradigm, this achievement serves as exemplar that unites prior work and points toward future work.[17] To acquire this status, the specific achievement must be unprecedented, attract a group of adherents to its purposes and methods, and be open-ended enough to leave more problems to solve.

Ordinarily, this achievement is an experiment whose results have far-reaching theoretical implications and is so suggestive that it cannot be ignored. Through further discourse, it comes to be associated with certain generalizations, metaphysical beliefs, and common values. This condition provides scientists with a vocabulary for articulating theoretical explanations, problems to be solved, and experiments to be done. Paradigms, then, organize scientific activity by providing conditions of social as well as intellectual cohesiveness.

They do not explain reality but instead suggest where and how to look for such explanations. They identify research questions as intimately related puzzles to solve and reassure scientists that the puzzles they are working on are all indeed parts of yet a larger puzzle that possesses a theoretical coherence.[18] Mid-range theories that link different smaller pieces of the puzzle togther are themselves linked together by grand theory as the frame of the total puzzle begins to become clear. Different groups of scientists may specialize to work on different aspects of this larger puzzle, but they will all be able to understand the foundations, meaning, and implications of one another's work. Kuhn calls this kind of practice normal science and claims that it characterizes mature sciences that display theoretical coherence and progress within a well developed paradigmatic framework.[19]

Kuhn's ideas about how science works and changes are not free of controversy.[20] Chief among them is Kuhn's use of the word *paradigm* in a number of different ways. According to one sympathetic critic the count in *The Structure of Scientific Revolutions* is twenty-one.[21] In this respect, it is equal to information in terms of ambiguity. A close second is its current popularity as a buzzword only remotely related to Kuhn's intent. This is nicely illustrated by a cartoon that portrays an executive returning to his office. His secretary is handing him a lengthy memo documenting the paradigm shifts that occurred while he was out. The popular use of the term does reflect the idea of challenge to received assumptions, but it also oversimplifies and tends to disguise routine conflicts of interest as something more profound.

A more serious controversy turns on the question of whether the concept is even appropriate to the social sciences. While warning against the idea that rigorous methodology is a "sufficient condition for scientific achievement," Abraham Kaplan argues that the methodology of the behavioral sciences is no different from that of any science and only subject matter separates them from other sciences. Crucial to any science is a commitment to the discovery and use of empirical evidence from which to draw inferences about nature.[22] Kaplan thinks that mathematics holds great promise for advancing the knowledge of the behavioral sciences,

> especially in making possible exact treatment of matters so long thought to be "intrinsically" incapable of it. I have no sympathy with principled, purposeful vagueness, even where it is not a cover for loose thinking. The argument that what we say about human behavior must be vague so as to mirror the richness of our subject-matter mistakes the nature of both description and explanation, as though these are to picture the reality, first as it appears and the second as it its underneath the appearance.[23]

For Kaplan, "science is a process, not just a product," and the former just might be more important than the latter.[24] This notion finds support in Karl Popper's insistence that a scientific proposition must be logically falsifiable, i.e., stated in such a way that its continency upon possible future discovery can be explicitly recognized.[25] For Popper, science is not about proving what we think to be true about nature. Rather it is about disproving what we claim to already know.

The process of science is more about investigating reality, grounded on certain logical rules of observation and inference, than about discovering truths that can be "knowledge." Given such a premise, there is no reason why human behavior cannot be examined scientifically. Human behavior is tangible, independent of our perception and consciousness of it, measurable in a physical and quantitative way, and it can be observed, predicted, and theorized about as with any other natural phenomenon.[26] What people are thinking about when they behave may not be easily discernable, possibly even beyond explanation, but the behavior itself and the conditions under which it occurs are quite material and entirely available to observation and analysis by scientific logic and method.[27] There is even some support for the notion that the behavioral approach in the social sciences satisfies Kuhn's criteria for a paradigm.[28] These ideas resonate with what Ellis identifies as the physical paradigm in information retrieval research. Information may be a human artifact, but there is no inherent reason why the study of its behavior cannot be an empirically grounded science.

On the other hand, Egon Guba challenges Kaplan's optimism regarding the application of science to the study of human behavior and asserts that behavior and its meaning cannot be so easily separated. The regularities of human behavior are themselves subject to at least two problems that raise doubts about whether it can serve as the object of scientific investigation. First, the meaning of these regularities is not constant. To observe one will not get us very far in the understanding of human conduct if the purpose of that conduct is variable and if its meaning is a social construction unique to a particular time, place, and group of people. In addition, any law of human behavior is potentially contradicted by the possibility that human beings are capable of willfully changing their nature, changing the rules by which they interact, or creating situations in which the rules do not apply.[29]

From Kuhn's perspective, the differences between the ideas of Kaplan and Guba represent conflicting ontological and epistemological assumptions that preclude designating the social sciences as "mature sciences." The manifestation of multiple, and in some cases apparently incommensurable, paradigms also precludes the practice of normal science. To use Kuhn's characterization, the social sciences are pre-paradigmatic, i.e., they have yet to become sciences, despite the perfectly reasonable application of scientific methodology to particular problems.

For Kuhn, however, a paradigm is about much more than just method. A paradigm includes theories, models, empirical laws, methodological rules, values, and metaphysical principles in a way that amounts to a "distinctive way of 'seeing' all the phenomena in its domain."[30] Paradigms identify sub-disciplines rather than whole disciplines; and while a so-called behavioral paradigm has a distinctive way of "seeing" a wide variety of social phenomena, it is characterized by its methods and its focus on behavior in general, rather than by its attention to any particular behavior. Herminio Martins puts it this way:

> Paradigms pertain to fields like the study of heat, optics, mechanics, etc.; there are not and cannot be paradigms of physics or chemistry. In other words paradigms are not discipline-wide but sub-disciplinary. Their span is likely to be coterminous with that of specialities: conversely, specialities will be paradigm-bonded social systems.[31]

Thus, a paradigm must be grounded in more than just a method. It must be based on a concrete scientific achievement that can serve as an exemplar for the investigation of a particular natural phenomenon.[32] The problem of assigning the status of normal science to the social sciences is not just a matter of conflicting methods and epistemologies. It is also a matter of the sheer number of phenomena that it is possible to include as appropriate objects of study within the social sciences, the variety of ways those phenomena may be conceived and described, and the continued existence of fundamental disagreements about both the phenomena that ought to be investigated and how to think about them.

Paradigms, Ambiguities, and the Social Sciences

In the social sciences there is little agreement about what their supposed paradigms are or have been. A particularly persistent discourse regarding fundamental questions of subject and method includes disagreements about how to conceive the nature of a paradigm in the social sciences.[33] In sociology, for example, the arbitrary and elusive character of the concept of "paradigm" allows two observers to identify twelve sets of paradigms. The paradigms within each set vary by theoretical focus, methodological approach, level of analysis, and the kind of social phenomena they address. Yet all of them plausibly characterize work in sociology and actually describe actual research.[34] It is not difficult to identify theoretical affinities among social scientists regarding problems of research. The social sciences, however, do not manifest communities of scholars bound together by a consensus regarding a widely recognized and specific scientific achievement, and so, fail to display what Kuhn argues is a necessary characteristic of normal science.[35]

Some critics accuse Kuhn of attempting to deprive science of its rationality, suggesting that he advances a claim that knowledge depends on what we want to know, but this criticism goes too far.[36] Rather, Kuhn is simply suggesting that what we can know depends on what we are prepared to know. A paradigm both invites and sets limits to investigation, and according to Kuhn, the outcome of this tension is both an intellectual and social phenomenon. It is difficult for people, including scientists, to step outside of the reality they seek to understand and view it dispassionately as merely a natural phenomenon, independent of their own lives. Methodology is used to obtain objective distance, but as Kaplan warns, it does not guarantee objectivity. If Kuhn is correct about the intrusion of the subjective world of the social into the sciences, then its intrusion on the practice of the social sciences must be even more influential.

It comes as no surprise that social scientists form neither communities or consensus based on allegiance to a given paradigm, given that the phenomena with which they work are themselves subject to profound disagreement regarding their existential nature and meaning. Typically, the names assigned to these phenomena must do double duty as serve both to describe and to create social reality. *Personality*, *family*, *race*, *market*, and *vote* are all examples of words that necessarily perform this double duty. They serve as fundamental theoretical and conceptualized objects of psychology, sociology, economics, and political science and as keywords whose essentially contested meanings focus the efforts from which human reality is made. This condition tends to impose a certain theoretical instability on the central concepts of the social sciences. Social scientists are often as divided among themselves about the nature and meaning of social phenomena, and the priorities with which they deserve attention, as are the ordinary people whose behavior create these phenomena.

Still, the social sciences clearly display characteristics of intellectual and social organization, and they manifest coherent "theory groups that help to organize research as a social practice." [37] These groups display at least two of the four elements Kuhn identifies as characteristics of the disciplinary matrix that explains the relative cohesiveness of a science.[38] They display metaphysical beliefs signified by basic assumptions regarding the nature of social existence and how it can be known, and they possess values that provide the critical grounds for separating good research from bad research. These characteristics contribute to setting disciplinary boundaries for the social sciences, but these disciplines do not quite live up to the demands of Kuhn's other criteria. The social sciences tend to lack symbolic generalizations, especially in the form of mathematical statements that express fundamental laws, and they lack exemplars, the concrete scientific achievements that signify a paradigm.

But if the social sciences lack symbolic generalizations and exemplars, how do we explain the organization, beliefs, values, and research that sustain them? What allows them to survive as coherent disciplines? Part of the answer to these questions lies in the fact that even if the social sciences are not as "organized" as the sciences, they are organized nonetheless. They are not entirely without symbolic generalizations, and they do possess examples, if not exemplars, that signify central ideas and ideals. Social theorists have invoked may images to describe "society." The latter has been described as a living system, a machine, a war, a legal order, a marketplace, a game, theater, and discourse.[39] Each of these images has served, and continues to serve as a powerful source of theoretical propositions and models in the social sciences. If these images cannot be associated with a specific scientific achievement, it is nevertheless not difficult to identify an example of each that continues to inform research.

The work of Talcott Parsons provides the groundwork for the functionalism and the idea of society as an integrated system. Auguste Compts' social physics informs the notion of society as a machine. Society as a war of all against all is found in Thomas Hobbes's, *Leviathan*, while John Locke's social contract theory supports the idea of society as a legal order. Adam Smith suggested in *The Wealth of Nations* that society could be seen as a marketplace, and the work of Anatol Rapoport and Thomas Schelling established game theory. Irving Goffman suggested that social interaction could be viewed as a form of theater, and the works of Michel Foucault and Jean Francois Lyotard examine society as discourse.

Admittedly, this is a very limited list, and we make no claim that any of these theorists signal a paradigm shift as did Copernicus, Newton, and Einstein. But there is no reason to assume that social sciences are less mature than "hard" sciences. Rather, they are less well organized. They display more internal divisions and disputes, and their progress is more difficult to define, let alone assess. The real key to the difference, however, is the social sciences are metaphorical; substitute the term *metaphor* for the term *paradigm*, and we see exactly how the social sciences can exercise theoretical and methodological discipline even in the face of ambiguous theoretical objects.

To begin with, let's return to Saussure. A science can be regarded as a syntagmatic process, or discourse, that slowly but completely changes its terms of discourse as the paradigmatic system that supports the process changes and substitutes new meanings for old ones. A social science is likewise a syntagmatic process, but its discursive phenomena are characterized by essentially contested meanings. Different paradigmatic systems plausibly, and more or less equally, provide different meanings for the terms of discourse in the social sciences. Their differences have to do with what aspect of the phenomena under investigation they choose to focus upon.

In the case of the social sciences the development of parallel syntagmatic processes are based on paradigmatically separate systems of thought and values. For example, each image, or metaphor, of society referred to earlier contains real truth value because each manifests aspects that can be empirically observed. Each metaphor of society embodies a logically coherent, empirically supportable paradigmatic system of ideas that poses research problems and suggests where and how to look for solutions to these problems. Yet because of the theoretical ambiguity of "society" as a concept, no one of these approaches can lay claim to dominant status. In addition, as a social science develops, its discourse will experience not so much a paradigm shift as a reinvention that spawns another parallel syntagmatic process. This is not a revolution based on a new paradigm but a matter of inventing new metaphors and applying them to the understanding of social phenomena.

Although metaphor's origins are in linguistics and poetics, George Lakoff and Mark Johnson elegantly explore its role in everyday speech: "the essence of metaphor is understanding and experiencing one thing in terms of another."[40] In other words, it allows us to make sense of ourselves and our world. A metaphor can also be described as an "operation of associative substitution" or a conceptual shift involving a substitution of meaning that operates along the paradigmatic axis of language[41] The use of metaphor invites us to understand the unknown by plausibly associating it with the meaning of something known. It allows us to connect two signifieds to the same signifier, to say "this is like that."

However, there are difficulties in doing so. One is that the use of *like* or *as* strictly speaking indicates a simile rather than a metaphor. To say that "this is like that" presumes that we already know what "that" is, and runs the risk of defining one indeterminate concept in terms of another. What must be granted for this idea to have force is that to say such a thing as "society is like war," or "information is like a physical object" is to say much more than the phrase literally denotes. We must accept that such phrases are deployed instead to connote meaning.

Still, the power of metaphorical thinking, and its application in the social sciences is unmistakable. "Society is war," and "information is a physical object" are metaphors that help us understand the general in terms of the particular. Lakoff calls these ontological metaphors. We use them to view intangible phenomena such as "events, activities, emotions, ideas, etc. as entities and substances."[42] They allow us to refer to these phenomena as causes and to act with respect to them. These metaphors express essential assumptions about the nature of existence, so that we may deal rationally with and make sense of experience—especially new experience.[43] We can even view paradigm as a special kind of metaphor. The puzzle-solving process of normal science is both paradigmatic and metaphorical in that a given phenomenon under investigation can be seen as analogous to the phenomenon represented by the exemplar. In other words, the unknown stands as a special case of the known.

Not all metaphors, however, are paradigms, and this is especially true in the social sciences, although, in two important ways, they are paradigmatic in nature. They allow research to be organized as a social and intellectual practice and then formed into disciplines and sub-disciplines that impose order on that practice. Because they are metaphors and not paradigms, however, they remain loose enough to cope with the disorganizing yet creative effects of the ambiguities of central theoretical objects.

A central organizing metaphor in a social science allows coherent thought, and even progress of a sort, precisely because it contributes to the social exercise of discipline within whatever group identifies with it. Accordingly, its use tends to privilege certain concepts, terms, definitions, rhetorical figures, and methods while excluding others.[44] As Lakoff and Johnson point out, however, concepts used to describe and construct social reality are themselves metaphorically constructed and culturally grounded.[45] Following Rousseau, we must remember that convention is never very far from the heart of human affairs. An unfortunate but necessary consequence of this condition is that equally compelling concepts, terms, and methods, ultimately essential to the understanding of human phenomena, may be excluded from some discourses and lines of research. This is the irony Neill intended to reveal when he claimed that some aspects of human reality must be excluded from investigation, or at least de-emphasized, in order to understand others.[46] Ambiguity is not so much resolved as eluded.

Elusion, however, is not necessarily a bad thing. By leaving the door open to uncertainty, we are encouraged to think pluralistically and resist the temptation to force reality to conform to theory. By employing metaphors instead of imposing paradigms, we can cope with disorganization and not lose the advantages of self-examination and understanding. Metaphors do this because they capture and express the ambiguities arising from theoretical objects whose phenomenological independence from human perception cannot be denied yet whose meaning is culturally and historically contingent. Metaphor allows us to bridge the gap of the object/subject categorical duality that characterizes human being, and make sense of our experience. The use of metaphor to talk about human experience recognizes that neither a strictly objective viewpoint nor a strictly subjective viewpoint is very helpful in the pursuit of understanding that experience.

Lakoff and Johnson argue that both positions miss the way we actually understand our world.[47] A strictly objective point of view denies that knowledge and truth are relative to cultural conceptual systems. No system of concepts regarding human nature can ever entirely escape its contingent origins and become neutral, absolute, and utterly disinterested. Human concept systems are metaphorical in nature and involve imaginative understanding. A strictly subjective point of view, on the other hand, denies that knowledge and truth, even the most imaginative and poetic, are given in terms of conceptual systems grounded on our successful functioning in physical and cultural environments and that imaginative responses to these environments that are not based on empirical observation and rationality are unlikely to be successful.[48]

The categorically dual nature of human reality does not imply a relativism that denies the possibility of knowledge, but it does suggest that it will be difficult to integrate alternative discourses regarding human reality. Truth is relative to conceptual systems we use to examine experience arising from daily interactions between ourselves and our world. Using metaphor to deploy the same signifier to apparently incommensurate signifieds can bring us tantalizing close sometimes to knowledge by creating plausible relations between the signifieds. "Reading as a cognitive process," for example, is a metaphor that brings us close to understanding the relation between "document" and "idea," yet even as each of these signs is reasonably well understood in its own right, the nature of the reality created by their combination still holds mysteries to be explored.

Using metaphors to understand the world, however, has some dangers. Clearly, metaphors do more than describe. They reflect the uncertainties and instabilities of our descriptions as they shape our conceptions and actions with regard to the phenomena they signify.[49] They can mislead and oversimplify, although this may be as much a matter of unexamined convention as it is deliberate effort. They are also historically contingent and vary across cultures. Still, the notion of metaphor can be very useful to explain and understand the structure and development of the social sciences, including information science.

Metaphors and Information Science

The concept of "information" provokes the same kind of difficulties that we encounter when we try to think in a disciplined manner about such concepts as "personality," "family," or "race." Like these concepts, it does double duty serving as a fundamental theoretical object of information science, and as a keyword whose essentially contested meaning focuses on activity that contributes to the making of an everyday world beyond and outside of the boundaries of information science as an intellectual discipline. This condition is nicely illustrated by Benjamin Disraeli's quip that "as a general rule, the most successful man in life is the man who has the best information," to say nothing of Cardinal Richelieu's habit of paying his spies out of his own pocket in order to ensure that he had the best information possible.[50]

The implication that there exists both good and bad information in turn raises questions regarding the criteria are applied in judgment. What are they? Whose are they? Would the criteria of science impose conclusions different from the criteria of politics or morality? What is at stake? Power, certainly, but applied to what end? Information is clearly a double-edged sword. Also at play is the commodity status of information. Its value is undoubtedly related to the knowledge it imparts, but can also be measured as cultural, political, and economic capital. Its objective reality cannot be denied, but what difference does it make? Because "information" is both an outcome of discovery and an instrument of creativity, its ambiguity as a theoretical concept is difficult to resolve.

The foregoing suggests that using the concept of paradigm to understanding the nature of information science as an intellectual discipline requires some caution. Ellis, for example, does not attempt to apply the concept to all of information science, but instead limits his focus to the sub-discipline of information retrieval research. Even so, he concedes that the so-called cognitive paradigm lacks an exemplar, a concrete scientific achievement that signifies the paradigm as a whole in the way that the Cranfield experiments of the 1950s signify the physical paradigm.[51] Given Kuhn's understanding of a paradigm, Ellis's application of the concept seems a bit strained at best. Nevertheless, the concept's power, and its continuing usefulness in debates over the proper way to talk about the practice of science, lies in its recognition that science is a social practice, subject to the same human variability that characterizes all social practices.

Given our excursion into metaphor, however, it is possible to take issue with Ellis's contention that information retrieval research is paradigmatic, but modifying it rather than dismissing it. His observation that information retrieval research is characterized by two distinct discourses is accurate and insightful. In fact, his notion of the categorical duality of information as a theoretical object, and his assertions regarding the categorical duality of information retrieval research, ironically and convincingly suggest an expansion of his general project and framework of analysis to the entire discipline of information science. His crucial insight that "information" possesses a dual nature that simultaneously embraces things and people, and his use of the word to signify both a physical object and a cognitive phenomena captures an essential dialectic that is pervasive in the discourse of information science. As we shall see, two broad discourses within information science have developed around this duality, each supported by individual practitioners, a structure of intellectual and economic capital, communication channels, and schools of library and information science. Each pursues research and articulates ideas and knowledge claims grounded on these two different interpretations of the sign "information."

The groups engaged in these discourses are not communities of scientists in a Kuhnian sense. They are far less organized, both socially and intellectually, than that identity merits. Nor is it fair to claim that either group represents a scientific paradigm at work or that they are engaged in a contest for dominant status in a scientific sense. Nevertheless, if the idea of "paradigm" does not exactly apply, something like two alternative camps of information science and information scientists can be identified. But here, too, metaphors rather than paradigms are at work.

At the core of the two camps that constitute a large part of information science practice, and provide the focus for the identities and loyalties of its practitioners, is a metaphor for "information" that each uses as its central organizing concept. For one, "information" is a physical object. This metaphor is premised on the idea that information can be conceived as an inert thing and natural object whose reality is independent of human perception and consciousness. Information appears in the world in objective forms that possess invariant characteristics and behave in regular and predictable ways regardless of the intentions of its producers, distributors, or users. For the other camp, "information" is a cognitive phenomenon. This metaphor is premised on the idea that information is a product of human consciousness and that it represents a structuring of thought for the purpose of intentional communication. From this perspective, information is regarded a subjective phenomenon whose nature and behavior depends on what people think about it. These metaphors perform many paradigmatic duties. They serve as sources of ontological assumptions about the nature of information and as sources of epistemological assumptions about how to come to know that nature. They organize and categorize research, control meaning, motivate actors, and articulate shared values.[52] They manifest symbolic generalizations of a sort, as well as metaphysical beliefs that provide the common ground for communication and group identity.

By substituting metaphor for paradigm we gain a number of advantages in our effort to understand information science. First, and foremost, it opens the door to tolerance of the ambiguity of "information" as a theoretical object. It need not be either objective phenomenon or subjective phenomenon. It can simultaneously be both, and members of both camps can draw on this simultaneity as needed. A second important advantage is that we can avoid the strict implications that follow from trying to use the notion "paradigm" to talk about research in information science. For example, we can admit that the "communities" identified with each metaphor are at best loose coalitions with great potential for easy border crossings. The two metaphors, and the camps identified by and with them, may be competing in one sense. Undoubtedly members of both camps are persuaded that the aspects of information on which they concentrate their attention are the most interesting, if not the most promising in terms of research breakthroughs and practical applications. But these camps are not in competition in the sense of excluding the other from practice according to Kuhn's notion of paradigmatic competition. Following the ideas of linguistics, it is more accurate to see these camps as constituting parallel processes of research, each driven by different systems of ideas that focus attention on different aspects of information as a theoretical object.

Finally, using "metaphor" rather than "paradigm" to investigate the structure and development of information science allows us to see how and why, given the creativity manifest at particular times and places, new metaphors grounded in new experiences and perceptions sometimes arise. These metaphors may not attract as much attention or as many adherents as older, more established metaphors, and they may borrow heavily from received ideas, but one need not deny their individuality nor force them into a theoretical framework defined exclusively by differences between the physical and cognitive metaphors. "Information" is a complex phenomenon related to multiple significations and so requires a variety of research approaches. There is every reason to believe that new approaches to and ideas about "information" will arise that do not neatly conform to either dominant camp. By abandoning the need to think about these experiments as potential paradigms, or to see them exclusively in terms of some existing framework of thought, we can allow them to stand on their own and contribute what they will to our knowledge of information. This may result in sacrificing disciplinary organization, but the trade-off may be worth the imaginative creative insights a new metaphor can supply.

The two metaphors identified here can appear to be set against one another in a categorical opposition. At the core of the difference between them are very different ideas about the nature of information as a theoretical object. Based on this difference, each metaphor poses very different kinds of research questions. At the heart of this contest over perspective is the question of whether information should be regarded primarily in its objective aspects or primarily in its subjective aspects. The essentially indeterminate relation between being informative and being informed, however, suggests that this division is more a matter of relative emphasis than sharp disagreement about the nature of information as a theoretical object. Nevertheless, disagreements about the nature of information are at play. The contest between the physical and cognitive metaphors begins with these disagreements, and they lead to disagreements about how to study the problems of information, as well as what can be known about information. Questions about the nature of existence raise issues of ontology. Questions about what we can know of what exists and how we know that raise issues of epistemology. Both sets of philosophical issues are closely related. What and how we can know about existence depends on the assumptions we make about its nature. The contest that concerns here us turns on several questions. What is the nature of the problem to be investigated? What kind of problem is it? What methods of investigation are appropriate? The difference in the answers provided by physical and cognitive metaphors respectively reveal the ontological and epistemological assumptions that constitute them as metaphors.

Despite their fundamental and profound differences, however, there are some important common threads bind these metaphors together. Chief among them is the notion that information science is a response to the human problem of establishing rational control over the expanding information universe. We must confront more texts and more documents, moving faster through larger channels of communication than could have been imagined even just a short time ago. The need to gain control over this universe, in order to make rational use of the record of knowledge and culture, manifests theoretical and practical aspects, problems of understanding and application, and ultimately the need to sort out what we think we know from what we really know. Access to the texts that record human knowledge and culture is the primary phenomenon of interest to information science, regardless of what other divisions may exist within the discipline.

Both metaphors are specifically concerned with texts as media of human communication. Specifically, both share the ontological assumption that information for information science demands a focus on texts as intentionally and purposefully structured collections of signs, regardless of the particular form a text may take or the medium of its communication. This structuring of texts as documents of human thought and communication is what allows information to be represented, organized, and retrieved. Certainly, the traditional document constituted by writing occupies a privileged position as the kind of text of primary concern to both metaphors, but by no means is this the only kind of text of interest. In addition, both metaphors share the epistemological assumption that texts-as-information are rationally intelligible, and that their behavior and use manifest regular and knowable patterns.

Both metaphors also reflect Shera's insight into why librarianship is still central in so many ways to information science. While librarianship remains focused on the social practices that constitute the library as an institution, and information science has moved beyond any particular institutional identification, at the heart of both disciplines is a concern for bibliographic control. Arguably, librarianship identified the problem that information science seeks to solve. This conclusion implies that the apparent differences between these two metaphors of "information" constitute a unity of opposites. Each has strengths that compensate for the other's weaknesses, and the research agendas guided by each displays a recognition of the other's value. Each needs the other. Research based on the physical metaphor is characterized by a relative homogeneity of purpose and method, offers a cumulation of results, and finds support from existing institutions and information service providers. It has not, however, generated an especially powerful explanatory theory sufficient to cope with the complexity and role of cognition as crucial aspect of becoming informed. In contrast, the cognitive metaphor suggests a way to solve some of the most troubling and difficult theoretical problems of information retrieval, but its complexity is not unlike a Gordian Knot in which understanding becomes lost in a tangle of over-explanation and little concrete result.

The unity of opposites constituted by the physical and cognitive metaphors can be illustrated by the situation of the librarian. Librarians appreciate that the classification systems and controlled vocabularies they employ everyday to organize the storage and retrieval of books in their collections are based on certain objective characteristics of books as informative objects. As a practical matter they use these tools to logically arrange books on shelves and to help people find those books. Without the benefits provided by logical means of organization and access, using books is difficult, if not impossible. This condition offers a confirmation of the everyday value of the physical metaphor. On the other hand, librarians are acutely aware of the fact that people do not generally come to libraries to get books. They come seeking the content and meaning of those books. The existence of books as physical objects whose inherent characteristics allow them to be organized in particular ways is not unrelated to the problems of finding the books, but is ultimately incidental to the purposes of people seeking information. This offers a confirmation of the everyday value of the cognitive metaphor. Both metaphors appear as necessary and yet insufficient. Neither can provide a fully coherent conceptual identity for the object they both call "information."

In addition, the image of two essentially opposing ideas contesting for the legitimacy to define the boundaries of information science repeats the long held western assumption of an essential opposition between object and subject, but just as this duality is open to challenge, so is the physical/cognitive duality in information science. Within recent years, some work in information science has been done using features of both metaphors which suggests that the relationship between the two may be more complex than it first appears. On one hand, we may be witnessing a genuine overcoming of disciplinary conflict as the relation between the object and subject of information science becomes better understood. On the other hand, it may be that the difference between the physical and cognitive perspectives was never quite as great as it once appeared. As we shall see later, there is reason to believe that not only do the physical and cognitive metaphors manifest a unity of opposites, they can also be regarded as two factions of a dominant theoretical approach to the study of information, and that within the fragmentation of information science, alternatives to both may be available.

Information science may not be identifiable as a "mature" science due to its lack of paradigms, the manifestation of multiple metaphors, and the continued presence of conflicting ontological and epistemological assumptions within its practice. There is reason to believe that the apparently insuperable problem at the center of information science regarding the ambiguity of information as a theoretical object may mean that the discipline is incapable of becoming a mature science. Nevertheless, in the meantime, it seems as if science is being practiced and that metaphors guiding that practice can be identified.

Endnotes

1. David Ellis, "The Physical and Cognitive Paradigms in Information Retrieval Research," *Journal of Documentation* 48 (March 1992): 60.
2. Ibid., 60-61.
3. Thomas Kuhn, *The Structure of Scientific Revolutions*, 2nd ed. (Chicago: University of Chicago Press, 1970).
4.Ibid., 68-69.
5 Ibid., 111-135.
6. Ibid., 23-34, 92-110.
7. Ibid., 10-22, 163.
8. Ellis, "The Physical and Cognitive Paradigms," 49.
9. Ibid., 53.
10. Paul Bouissac, "Paradigm," in *Encyclopedia of Semiotics*, ed. Paul Bouissac. (Oxford: Oxford University Press, 1998), 460-461.
11. Ferdinand De Saussure, *Course in General Linguistics,* ed. Charles Bally and Albert Sechehaye, trans. Wade Baskin (New York: Philosophical Library, 1959), 122-127.
12. J. van Marle, "Paradigms," in *The Encyclopedia of Language and Linguistics,* ed. R. E. Asher vol. 6 (Oxford: Pergamon Press, 1994), 2927-2930.
13. Ibid., 2928; and Bouissac, "Paradigm," 461.
14. van Marle, "Paradigms," 2929.
15. Bouissac, "Paradigm," 461.
16. R. Jakobson, "Linguistics and Poetics," in *Style and Language*, ed. T. A. Sebeok (Cambridge, Mass.: MIT Press, 1960), 350-377.
17. Ellis, "Physical and Cognitive Paradigms," 46-48.
18. Kuhn, *Structure*, 35-42.
19. Ibid.; and Ellis, "Physical and Cognitive Paradigms," 48-49, 56-58.

20. D. C. Stove, *Popper and After : Four Modern Irrationalists* (New York: Pergamon Press, 1982), and *Paradigms and Revolutions : Appraisals and Applications of Thomas Kuhn's Philosophy of Science* ed. Gary Gutting (Notre Dame, Ind.: University of Notre Dame Press, 1980).
21. Margaret Masterman, "The Nature of a Paradigm," in *Criticism and the Growth of Knowledge*, ed. I. Lakatos and A. Musgrave (Cambridge: Cambridge University Press, 1970), 59-91.
22. Abraham Kaplan, *The Conduct of Inquiry: Methodology for the Behavioral Sciences* (Scranton, Penn.: Chandler Publishing Company, 1964), 25, 30, 32, 34-46.
23. Ibid., 409.
24. Ibid.
25. Karl Popper, *The Logic of Discovery* (New York: Basic Books, Inc.), 40-41.
26. Kaplan, *Conduct*, 136-144, 206-214, 322-326.
27. Heinz Eulau, *The Behavioral Persuasion in Politics* (New York: Random House, 1963); and Arthur L. Stinchombe, *Constructing Social Theories* (New York: Harcourt, Brace & World, Inc., 1968).
28. Sheldon Wolin, "Paradigms and Political Theories," in *Paradigms and Revolutions*, 181.
29. Yvonna S. Lincoln and Egon G. Guba, *Naturalistic Inquiry* (Newbury Park, Calif.: Sage Publications, 1985).
30. Gary Gutting, "Introduction," in *Paradigms and Revolutions*, 12.
31. Herminio Martins, "The Kuhnian 'Revolution' and Its Implications for Sociology," in *Imagination and Precision in the Social Sciences,* ed. T .J. Nossiter, A. H. Hanson, and Stein Rokkan (London: Faber and Faber, 1972), 19.
32. Gutting, "Introduction," in *Paradigms and Revolutions,* 14.
33. Ibid., 13.
34. Douglas Lee Eckberg and Lester Hill, "The Paradigm Concept and Sociology: A Critical Review," in *Paradigms and Revolutions,* 122, 124, 132.
35. Gutting, "Introduction," 13; and Eckberg and Hill, "The Paradigm Concept," in *Paradigms and Revolutions,* 124; Nicholas C. Mullins, *Theories and Theory Groups in Contemporary American Sociology* (New York: Harper and Row, 1973).
36. Imre Lakatos, *The Methodology of Scientific Research Programmes: Philosophical Papers*, Vol. 1, ed. John Worrall and Gregory Curries (New York: Cambridge University Press, 1987), 90-93.
37. Nicholas C. Mullins, *Theories and Theory Groups in Contemporary American Sociology* (New York: Harper & Row, 1973).
38. Kuhn, *Structure*, 182.
39. Daniel Rigney, *The Metaphorical Society: An Invitation to Social Theory* (Lanham, Md.: Rowman & Littlefield Publishers, Inc., 2001), 6-10.
40. George Lakoff and Mark Johnson, *The Metaphors We live By* (Chicago: University of Chicago Press, 1980), 5.
41. Jere Paul Surber, *Culture and Critique: An Introduction to the Critical Discourses of Cultural Studies* (Boulder, Colo.: Westview Press, 1998), 166.

42. Lakoff and Johnson, *Metaphorical Society*, 25.
43. Ibid., 26.
44. Surber, *Culture and Critique*, 188.
45. Lakoff and Johnson, *Metaphorical Society*, 8-9.
46. S.D. Neill, "The Dilemma of the Subjective in Information Retrieval," *Journal of Documentation* 43 (September 1987):193-211.
47. Lakoff and Johnson, *Metaphorical Society*, 194.
48. Ibid., 185-194.
49. Rigney, *Metaphorical Society*, 2-4.
50. *The New Book of Unusual Quotations*, sel. and ed. Rudolf Flesch (New York: Harper and Row, 1966), 180; and Ronald Seth, *The Encyclopedia of Espionage* (London: New English Library, 1972), 526.
51. Ellis, "The Physical and Cognitive Paradigms," 50, 53.
52. Ibid., 201.

3

The Physical Metaphor

Without doubt, information appears to us in a material form. For a variety of reasons, data, information, knowledge, and wisdom are recorded as texts, and these texts, in the form of writing, images, sounds, or any other medium apprehensible by the senses, can be described in a language of physical things. As material objects, they can then be collected, organized, and retrieved for use. This straightforward manifestation of information in a material form provides the foundation of the physical metaphor in information science. At the center of this metaphor is the idea that information can be regarded as a fundamental and universal phenomenon similar to matter and energy. Just as energy is manifest in a variety of forms such as heat, light, and electricity, so is information in such forms as knowledge, news, and data.

Although information reveals itself in a variety of settings and attributes, essential objective aspects of information can be described abstractly and independently of the particular form in which it appears. This characteristic allows it to be manipulated in logical ways and suggests the existence of fundamental and invariant laws which might govern its behavior. Here too, just as energy can stored, transferred, and made to do work on the basis of certain fundamental laws regarding its nature and behavior, information can be stored, retrieved, and communicated on a similar basis. If true, this condition implies that information itself is a kind of physical, tangible object that can be observed and, like other such objects, can be controlled by means of a technology. The growth in quantity of information and the complexity of its structures pose difficult control problems but the nature of information as an object suggests that these problems can be reduced to general and elementary operations.[1] This idea of information-as-thing provides the foundation for a number of solutions to the theoretical and practical problems of information storage and retrieval.[2]

Conceiving information as a thing can violate our intuitive expectations about its nature. Conventionally we tend to think of information in terms of meaningful content—a fact, an explanation, an evaluation, a thesis, or an idea. Rather than being a thing in itself, we tend to regard information as an attribute of things. Using the word *information* to signify a thing in itself obliges us to think about information in a very particular way. The physical metaphor assumes that a natural, objective reality exists independently of human consciousness and perception and that information is a phenomenon of this reality. Its nature and behavior awaits to be discovered. From this perspective, information would possess regular and predictable characteristics and behavior like any other natural object, which in turn we can manipulate and control. In this respect, information is no different from the elements of the periodic table of elements or the planets and their movements. The more we know about it, the greater and more precise our control of its behavior will be.

An objection might be raised that to detach unique meaning from a particular bit of information in order to better control information in general defeats the purpose of control. After all, a user of information is usually interested in its particular not its general characteristics. This objection is a reminder of the ambiguity and categorical duality of information as a theoretical object, and it leads back to the conventional expectation that the detachment of meaning from information cannot be sustained. "Information" must signify both an object and its meaning. But if this is the case, then of what value is an approach to the study and understanding of information that detaches meaning from information and the focus of which is on information as a natural object? How does this metaphor retain its power to guide research and solve practical problems?

Answers to these questions must factor in two considerations. First, the physical metaphor does not entirely abandon the issue of meaning. Second, it represents the need to reduce complex phenomena to essential elements in order to study such phenomena, and even though knowledge may fall short of complete understanding, research into the smaller aspects of reality may still yield tangible and useful results.

The need to simplify complex phenomena in order to study them is a general issue of science. In physics, for example, it is necessary to understand a very basic idea (e.g., that velocity is a function of a relation between distance and time) before it is possible to think about the nature of the relations between energy, mass, and the speed of light. Similarly, solving a basic problem of information science, the need to organize and make accessible the record of human knowledge and culture must precede a consideration do so in a way that actually leads to the growth of knowledge. Several issues have to be sorted out before we can approach a solution to the latter problem. What are the important and essential elements of this problem? What needs to be known to solve it and in what priority should studies be conducted? What can be known about the elements of the problem and how can we know about them? How can we be confident about our knowledge? These are fundamental questions of epistemology, and their answers contribute to the formation of metaphors regarding the nature of information. A science, whether about information or anything else, must begin with some notion of what ought to be regarded as the essential elements of the problem it wants to solve and the reality it seeks to study and understand.

The physical metaphor of information science begins with observations of the general and invariant characteristics of informative objects, most commonly texts, and then seeks to know how and why these objects behave as they do under different conditions. The early efforts of librarians and documentalists to reduce the problem of information to its essential elements, for example, led to posing questions about and seeking responses to the classic challenges of bibliographic control. They were trying to identify the defining characteristics of written texts in order to organize them for retrieval. The physical metaphor's approach to the problem of information is illustrated by asking questions about texts, regardless of the medium in which text appears and is communicated. What characteristics does the text possess that are invariant, important, necessary, and sufficient for the purpose of comparing it to other texts, of assessing its similarities and differences regarding other texts, and organizing it along with other texts in order to make it accessible when desired? What kind of system of organization and access can be built based upon these characteristics? How can systems of organization and access be compared in order to yield empirical and confident knowledge about which system represents the best means of retrieving desired texts? Central to each question is the assumption that texts possess certain characteristics that makes them organizable and retrievable in logical and predictable ways.

A particular kind of text illustrates this point beautifully. Given that there are always exceptions to the rule, books have in general at least four distinct characteristics. A book has an author (whether an individual, a corporate body, or anonymous), a title, and a subject; and it is an independent intellectual entity conventionally of some considerable length. While none of the these characteristics are absolute and all are subject to a degree of judgment, in practice, the first two (author and title) help us to uniquely identify and describe the book, the third (subject) is crucial to representing its aboutness, while the last allows us to distinguish it from some other form of written text such as an article. The difficulties involved in accomplishing these tasks are not trivial. The common tools used to do this, for example the *Anglo-American Cataloguing Rules* and the *Library of Congress Subject Headings*, are themselves texts whose complexity reveal the subtleties and difficulties that these tasks may imply. Describing a book is a great deal more difficult than may at first appear.

Nevertheless, the crucial point here is that a book will possess certain characteristics which define it as a book—and as a particular book—different from other books. In addition, these characteristics are of a given nature and they do not change, regardless of what we think about them. If we believe a book to have a different title from the one it actually has, then we are wrong. The subject of a book clearly raises problems of ambiguity, but if we believe a book is about furniture when it is actually about animals, then again we are wrong. The book remains what it is and it remains the same regardless of what we may think of it. If the meaning of the book is to come under consideration at all, the only valid place to look for it is in the thoughts of the author rather than in the mind of its reader. Even then, we may be on shaky ground since the author may have been unable to truly express what he or she was really thinking. From this perspective, information must be regarded only as what is physically present in the words and sentences that make up a text. It cannot be reliably found in the mind of either the creator or user of a text.[3]

For the purposes of a science of information, and more particularly a science of information retrieval, the fact that books can be characterized by author, title, and subject is far more important in general than the specific content of any of these categories for any particular book. Because these characteristics may be regarded as the invariant characteristics of all books and any book, we can create author, title, and subject index files which organize books for access. The printed card catalog, although now taken for granted and nearing complete obsolescence, represents a powerful organizing and access technology based on the nature of books as physical, tangible objects. In a real way, author, title, and subject can be regarded as crucial characteristics of books that contribute to an operational definition of the general concept "book." The object, "book" in this case, is defined in terms of its observable characteristics. The card catalog exploits and represents these characteristics in the form of physical document surrogates, for the purpose of bibliographic control.

It is the operational definition of natural objects and phenomena in a language of physical things and the means by which their characteristics and attributes are distinguished and measured that forms the building blocks of a science. The next step involves the study of how these objects and phenomena tend to behave. But inevitably, a few items will probably not behave as expected.

Their behavior cannot be predicted, or they do not fit the system designed to explain and control their behavior. Given a collection of texts from which we desire to retrieve particular texts, there will always be some which are relevant to our desire but not retrieved. However, if we can design our systems according to how the objects we wish to control are most likely to behave under given circumstances, then we maximize the probability that those systems will effectively and efficiently control them in ways that we desire. Control of information is based on the assumption that we regard texts as natural objects with characteristics and behavior independent of our consciousness of them. Herein lie the roots of the physical metaphor in information science.

The Nature of Information and the Physical Metaphor

To fully understand the project of the physical metaphor we must begin with the notion that there are tangible objects and expressions with the quality of being informative. This quality, while characteristic of the formal structure of that object in itself, does not depend on whether the object is perceived to be informative nor on whether it actually informs anyone.[4] This condition resembles the potential energy in a tightly wound spring. To do work, for example to drive a child's toy across the floor, the spring must be released, allowing the transformation of potential energy into kinetic energy. This potential may not be used—the toy may not be set in motion—but potential energy remains an inherent, objective, formal, and measurable characteristic of the spring.

Texts belong to a class of objects and expressions that possess an inherent potential to inform that exists independently of their use. This potential to inform is released whenever someone reads a text. The fact that some texts are not perceived as informative or used to inform does not negate or diminish their informative quality. Unlike the spring, which over time inevitably loses tension to stress, the informative quality of a text is constant and unchanging and constitutes the aboutness of that text. However, this idea must be treated with care. A text, relative to newer texts, may become obsolete, its information subsumed or superceded by other information in those other texts, but this is a matter of use. Nevertheless, the intrinsic content of a text does not change, and neither does its aboutness. The slow release of tension in a spring might be compared to a text that unwrites itself, a most unlikely event. In addition, the use of an informative object does not negate or diminish its quality of being informative. A single such object may be used an infinite number of times in a variety of ways. In other words, an object or expression that possesses the quality of being informative also possesses an intrinsic aboutness regarding its content that is independent of its use or meaning.[5]

This intrinsic aboutness or, more conventionally, the subject of information-as-thing varies from thing to thing; but for each unique information-thing, it is an invariant and objective characteristic that allows us to distinguish among them in the set of all information-things. The particular quality of an object or expression as information, or as potentially informative, is a characteristic of the object or expression itself, which in turn makes communication possible. By serving as a constant and unchanging reference point, it allows two speakers of different languages to point at an apple, make an eating gesture, and understand that the object is food. On a more sophisticated level, it is precisely this quality of words that allows us to make sense of sentences we have never before heard, and to logically store, organize, and retrieve information as needed by means of tangible, objective, and predictable processes. The paradigmatic quality of language that allows one word to substitute for another is essentially related to the informative quality of any word and any text, and it makes possible representation for retrieval.

In this concept of information there is clearly a place for meaning, but this place is limited by the claim that meaning itself is an intangible and subjective phenomenon, unique to each reader of a text, and that it can play no role in information retrieval.[6] From the perspective of the physical metaphor, acts and outcomes of informing and becoming informed are essentially subjective phenomena that occur within the mind of a user of information and are neither observable or describable in a language of physical things. We might observe behavior that follows becoming informed and draw inferences from those observations about the influence of information. This activity does have value for information retrieval, but the physical metaphor of information science insists that systems of information retrieval cannot be based on a phenomenon as subjective, idiosyncratic, and arbitrary as meaning.[7]

If, for example, one of the aforementioned speakers believed the apple to be a forbidden fruit, then the gesture of eating one might strike him as heresy. Only with some difficulty might the two speakers establish that even though only one of them could actually eat it, an apple is nevertheless an edible object. In other words, they might agree on the objective, intrinsic properties of the fruit even as they disagree on its cultural meaning. Shared meaning is necessary to communication, and it can be established only through inter-subjective interpretation of informative objects and expressions, but if information did not exist as a thing, in a thing, prior to communication, there would be no grounds for arriving at a shared meaning; there would be no common, objective reference point around which to create a shared experience. Communication can begin only after the objective properties of informative objects have been recognized.

From the perspective of the physical metaphor, the intangible and subjective nature of information is beyond knowing, and as such is beyond control, prediction, and manipulation. The interpretation and creation of meaning from information, while central to the purpose of retrieving information, are nevertheless acts and outcomes that lie beyond the mechanism and logic of retrieving of information. They are phenomena over which an information scientist can have no influence or control. Information retrieval *systems*, regardless of the kind of technology that supports them, require that the objects and expressions to be stored, organized, and retrieved possess certain constant and unchanging characteristics by which they can be distinguished, grouped, and classified. Whether a print-based card catalog or a computer-based search engine, an information retrieval system must be based on logical, systematic rules for drawing inferences from and about the inherent and invariant characteristics of stored information in order to retrieve it. Once retrieved, its use is relative to its users' purposes. Information retrieval systems cannot operate on intangible and subjective phenomena.[8]

The physical metaphor of information science is based on a very practical observation that can be illustrated by some straightforward examples. Libraries do not organize knowledge. They organize books on shelves and they organize representations of these books in catalogs. Online databases provide access not to knowledge but to index terms and by means of this access they retrieve document citations or, in some cases, full text of documents. In each case, what is stored, organized, retrieved, and processed is information-as-thing, a material, tangible object, a text. The intangible and subjective nature of information is transformed into objects or surrogates that represent it, and information retrieval systems operate on and with these representations. Specifically, they operate on and with the intrinsic aboutness of informative objects and expressions by translating that quality into representations of their content, which we may think of as information-as-knowledge. By treating information as a thing, the physical metaphor poses a limited and straightforward but far from easy puzzle to solve. What do we need to know about information in order to design information retrieval systems that perform at optimal effectiveness and efficiency? How can we ensure that these systems successfully predict and retrieve the texts which will most likely satisfy a user's request, based on a match between representations of the intrinsic informative qualities of the stored texts and the request?[9]

It is clear that all sorts of objects and expressions may be treated as informative objects in this manner. A wide variety of texts, natural and human-made objects, and even people can represent knowledge and be represented, in turn, as potentially informative objects. Each is a tangible and objective phenomenon, by its nature possessing intrinsically informative qualities. Arguably, information science need not limit its research to a consideration of only those phenomena that are intentionally informative.[10] Nevertheless information science, guided by the physical metaphor, has traditionally focused on four objects: statements (documents) made by originators of knowledge or communication, requests (queries) made to information retrieval systems designed to store and organize statements of originators, translated versions of these objects (document and query representations),and system output (retrieved documents). Resolution of uncertainty, learning, acquiring understanding and knowledge, and interpretation of meaning are all perhaps at the source of the motive to use an information retrieval system, but these concepts relate to mental states which are not accessible to observation, and from the perspective of the physical metaphor they cannot serve as a starting point for a science of information.[11]

It is too simple but perhaps not entirely unfair to say that the physical metaphor is sustained by a simple operational definition of information as documents and their representations. Similarly, the word *text* is often used to connote the variability of objects and expressions that can be considered to be informative, though in practice the physical metaphor has always been primarily concerned with the bibliographic control of documents, specifically those that constitute scientific literature and discourse.[12] Even with the advent of electronic publishing and the development of the Internet, it is still fair to say that the document is still the physical metaphor for information. In turn, reducing the concept of information to a matter of documents and their representation allows the work conducted in the context of the physical metaphor to proceed in a manner free of a number of the theoretical ambiguities and confusing intuitive notions of "information."

It is vital to remember that information, in the physical form of the representation of knowledge, is the only external, observable element in the communication process. Therefore, if information science is to be an empirically grounded, experimental science, it must focus on such observable elements as can be isolated within the system in which they typically interact. In turn, each element must be capable of independent manipulation. In this way, the physical metaphor allows and encourages the intellectual possibility of a discourse that proceeds from the study of simple phenomena to the study of related but more complex information phenomena, such as information-seeking behavior. Through examining the nature and behavior of documents then, we can begin to understand the consequences of their use and the knowledge they contain. Perhaps the acquisition of knowledge is the "real" goal of the users of information retrieval systems, but final effects of using information are the outcome of reading and thought, two phenomena that are beyond observation. Nevertheless, the relations between content and meaning, and the effects of information on recipients of information are clearly phenomena of interest to information science, and it may be possible to infer these effects from the observable behavior with which they are associated.

The documents stored in information retrieval systems and the queries posed to retrieve them, however, are observable phenomena, and from this perspective, information retrieval can be treated as if it is, in effect, document retrieval.[13] If information science limits itself to the issue of improving the means of representing documents and queries for the purpose of improving the control and retrieval of documents, then it sets for itself an achievable project. From the perspective of the physical metaphor, this project involves discovering ways to improve the means of representing documents and queries and to develop more effective techniques for maximizing the accessibility of documents to users and maximizing the exposure of users to documents.[14] Use of the physical metaphor allows us to see what happens inside the system—where and when queries meet documents.

The physical metaphor directs attention to the features of queries and documents that can be represented, how can they most accurately and adequately be represented, and how those representations can be manipulated to produce desired results. The image of striving to design and engineer a machine whose working approaches perfection is not accidental. This image arises from the attention of the physical metaphor on classification, indexing, and processing of language by means of computer technology.[15] Experimental design and controlled experiments then yield quantitative measures of performance so that pragmatic decisions may be made about the feasibility of particular design principles as well as the advisability of investment in particular information retrieval system developments based on those principles.[16]

The Physical Metaphor and Talk About "Information"

The nature of "information" as a theoretical object as well as the means of its metaphorical construction is revealed as much in how it is discussed as it is by its formal definition. One common view of information science, for example, encompasses "all phases of the information transfer process" with the goal of standardizing the efficiency of this process.[17, 18] The computer suggests itself as the ideal machine for accomplishing this goal. Indeed, to the extent that the process can be automated,

> it should also be obvious that the provision of better service to the user, and the intelligent application of modern technology are not mutually exclusive enterprises, and the computer—as yet unable to reply to its critics without assistance—is usually made the scapegoat when something goes wrong. In fact, if it were not for the beneficial effects of the computer in modern society, we would be living in a Kafkaesque nightmare, replete with error-prone and degrading tasks that humans would be forced to do. There is also the very real possibility that we would no longer manage by using strictly manual procedures.[19]

The result is an increased ability to avoid errors, inconsistencies, and the unknowable and sometimes unreliable process of subjective human judgment regarding the control of information.[20]

To achieve standardization and subsequently automation, information science seeks a "logical, mathematical method" for the analysis and control of textual aboutness. In this context, focus shifts to devices of vocabulary control, such as the frequency of occurrence and the location of words in a text. Their associations with other words and sentences, including proximity, syntactical, and semantic associations, all become objects of theoretical attention. This research is aimed at the development of "devices" of vocabulary control that contribute to our control of information.

> With respect to information storage and retrieval specifically, we should note that the purpose of all vocabulary control devices—whether they are subject heading lists, thesauri, or any of the specialized classification schemes used for indexing—is to promote order by providing greater specificity and minimizing ambiguity.[21]

It is assumed here that "information is stored in documents," and that with the aid of appropriate devices, this information can be extracted.[22] The practical goal of this work is to achieve "optimal output," in other words to retrieve a number of documents that is not so large as to preclude their rational examination nor so small that "information" relevant to a request is not retrieved.[23] From this perspective, "it is useful to think of collections of information centers and libraries or, for that matter, of any given data bases as sets of items" (i.e., documents).[24] Thinking about "information" in this way allows us to apply set theory and Boolean logic to problems of information control.

A set is a collection of items that meets three criteria. First, membership of an item in a set can be determined by an unambiguous yes/no judgment regarding its possession of a qualifying and defining attribute; second, each item must be unique; and third, the order of each item in the set is irrelevant to its qualifications for membership. If we substitute "documents" for "item" then we can apply the principles of Boolean logic to the control of documents.

Now a common feature of online databases, library catalogs, and Internet search engines, Boolean search techniques are simply a matter of combining sets. They can be applied to either controlled vocabulary searches, where strategies are composed by selecting terms that have been assigned to documents from a sanctioned list of subject descriptor words, or free-text searches where searchers are free to select any words they believe might be present in the documents to be searched and that represent the subjects of the desired documents.

For purposes of illustration, let us search a database with various sets of documents, at least one of which was created by assigning to its members subject term A, and another of which was created by assigning to its members subject term B. If we were interested in documents based on the attribute of being described by either term A or B, we could simply retrieve both of these sets and begin looking through them, but Boolean logic offers a much more efficient way of doing so. The three basic

Boolean operators are *and*, *or*, and *not*. Generally speaking, the *and* operator is used to narrow the focus of a search, and provide greater specificity of output. The *or* operator is used to expand the focus of a search, and the *not* operator is used to exclude unwanted textual elements, ideas, and subjects that might distract from the searcher's purposes or confuse the issue under investigation.

Entering the statement "A and B" will retrieve only those documents that have been assigned both subject terms, creating a set much smaller than the two from which it came. The Venn diagram in Figure 3.1 illustrates this outcome.

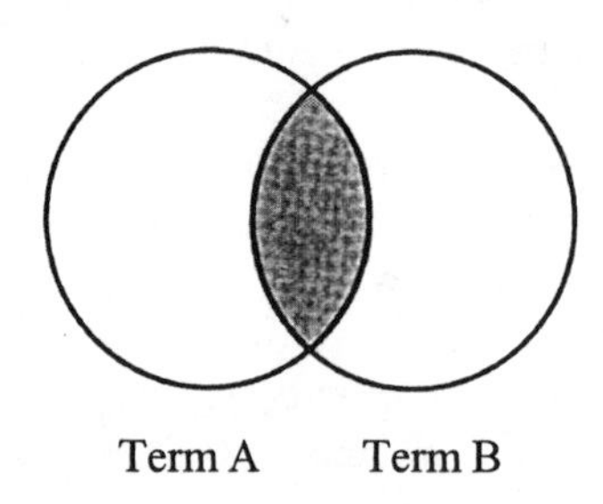

Figure 3.1 A and B, Shaded Area

Entering the statement "A or B" will retrieve documents that are assigned either term A or B, creating a much larger set. The Venn diagram in Figure 3.2 illustrates this outcome.

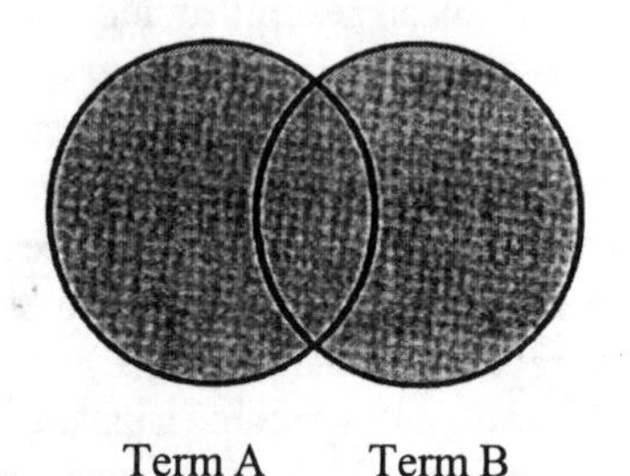

Figure 3.2 A or B, Shaded Area

Entering the statement "A not B" will retrieve documents that were assigned term A, but will exclude those term A documents that also have been assigned term B, creating a set that has more documents than in the "A and B" set, but fewer than in "A or B" set. The Venn diagram in Figure 3.3 illustrates this outcome.

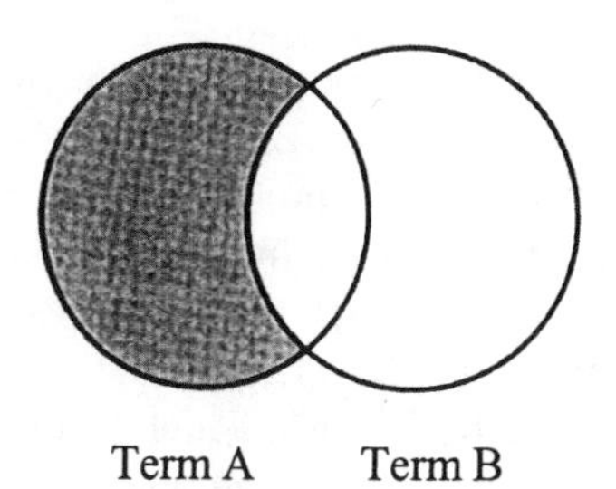

Figure 3.3 A not B, Shaded Area

All three operators can be combined with a number of different search terms to create very complex search strategies and very specific sets of retrieval output, for example "[(A and B) or C] not D." This statement will first create a small set of documents that have been assigned both terms A and B. Then it will expand that set by adding to it all documents assigned term C. Finally, it will exclude any documents that meet these criteria but have also been assigned term D. While there is some controversy surrounding the effectiveness of complex search strategies, Boolean logic does allow relatively easy manipulation and control of the aboutness of texts in large collections.

What we have here is a kind of talk about "information," and while it lacks a formal definition of information, it is nonetheless concerned with an understanding of the nature of information as a theoretical object. The language employed is not accidental. Its references to transfer processes, logical and mathematical methods, devices, optimal output, the role of technology (especially computers) order, and efficiency all speak of a particular way of thinking about "information" that is present but not always explicitly articulated. It is a language of physical things.

It speaks to, but not of, communication. It implies that communication is a process of transferring discrete units of information from source to receiver, governed by a discoverable mechanics.[25] The nature of information lies hidden in the background of this talk as a given, concealing the implicit signification of its discourse. A metaphor at work here that precedes the practice the talk describes.

Such concealment of this metaphor does not represent a deliberate effort to deceive. Instead, it illustrates the way the physical metaphor posits information as a thing and specifically as a natural object existing in the world independent of our perception or consciousness of it. This object possesses knowable and, in this context, even measurable invariant attributes that allows it to be rationally manipulated and controlled, especially when it takes the form of a document. The metaphor's work is observably evident in the talk about information and set theory, where libraries and information centers are equated with databases and reduced to sets of items. In this way, institutions are stripped of their cultural context and meaning, transformed into devices, and reconstituted as systems whose nature and dynamics can be understood by the same means we employ to understand any natural system. The most interesting aspect of this transformation is that it is not done with the intention of deliberate oversimplification, but through what we can only call oblivion. The language is quite matter-of-fact in its insistence that information retrieval is a mechanical rather than a social act. From this perspective, a library is just another kind of database.

Documents too are transformed into items and things with knowable, finite, and determinant attributes not subject to the vagaries imposed by the social and cultural contexts of their production and use. The "yes or no" quality of judgements regarding the membership of an information item in a particular set is not problematic. "Maybe" is not an option, because the informative attributes of documents in a database are assumed to be objective and discoverable, and each document, as well as the information it conveys, is discrete, self-contained, and independent of any other item in the set. For the purpose of retrieval, intertextual relations between items are implied by their identification with the same set defining index term such that the identity of each item as a member of a given set implies a relationship, even as it stands on its own as a unique informative object. It is as if the flow of a river could be broken into segments and each segment discretely identified.

Conclusion

This talk describes "information" as a thing, like an object of nature, but at the same time it cannot quite escape the ambiguity it seeks to resolve. The assertion that information is stored in documents is telling. Other words that plausibly describe the relation of information to a document come to mind. Words like *manifest* and *immanent*, for example, might be acceptable as plausible synonyms for *stored*, yet they suggest that information might be an intangible, ephemeral, and non-physical phenomenon. *Stored*, on the other hand, suggests the physical placement of a thing into an container, but to say that information is *in* a document raises certain problems. Does this mean that information can literally be equated with the document that contains it? Or is there a subtlety that eludes this concept of information?

The physical metaphor reduces the problem of information to an issue of how to control informative objects. The goal of this research is to achieve a depth and specificity of representation that enhances the organization of information and increases the probability of retrieving requested documents and presumably information. Overall, it has been quite successful. Issues of the effects of information on users are not regarded as unimportant, but the unobservable quality of these effects are regarded as problematic for a science of information retrieval. The elegant simplicity of the physical metaphor when combined with the rapid development of information technology encourages the development of increasingly sophisticated retrieval systems. This metaphor continues to identify and describe puzzles for research in information science. In the next chapter we examine an idea, a method, and an experiment—each of which represents the metaphor and embodies information-as-thing as the central and essential theoretical object of information science.

Endnotes

1. Klaus Otten and Anthony Debons, "Towards a Metascience of Information: Informatology," *Journal of the American Society for Information Science* 21 (January/February 1970): 90.
2. Michael K. Buckland, "Information as Thing," *Journal of the American Society for Information Science* 42 (June 1991): 351-352.
3. J. Farradane, "The Nature of Information," *Journal of Information Science* 1 (April 1979): 14.
4. Buckland, "Information as Thing," 352-353.
5. Clare Beghtol, "Bibliographic Classification Theory and Text Linguistics: Aboutness Analysis, Intertextuality and the Cognitive Act of Classifying Documents," *Journal of Documentation* 42 (June 1986): 84-86.
6 J. Farradane, "Nature of Information," 13-14.
7. M. E. Maron, "On Indexing, Retrieval, and the Meaning of About," *Journal of the American Society for Information Science* 28 (January 1977): 40-41.
8. Buckland, "Information as Thing," 352.
9. Donald B. Cleveland and Ana D. Cleveland, *Introduction to Indexing and Abstracting* (Littleton, Colo.: Libraries Unlimited, 1983), 19-26.

10. Buckland, "Information as Thing," 354.
11. Farradane, "Nature of Information," 15-17.
12. Alvin M. Schrader, "In Search of a Name: Information Science and Its Conceptual Antecedents," *Library and Information Science Research* 6 (September 1984): 230-239.
13. Charles H. Davis and James E. Rush, *Guide to Information Science* (Westport, Conn.: Greenwood Press, 1979), 61-66.
14. F. W. Lancaster, *The Measurement and Evaluation of Library Services* (Washington, D.C.: Information Resources Press, 1977), 11-13.
15. Peter Ingwersen, *Information Retrieval Interaction* (London: Taylor Graham, 1992), 62.
16. Stephen E. Robertson, "The Methodology of the Information Retrieval Experiment," in *Information Retrieval Experiment,* ed. Karen Sparck Jones (London, Boston: Butterworths, 1981), 10-11.
17. Davis and Rush, *Guide to Information Science*, 3.
18. Ibid., 15-16.
19. Ibid., 7.
20. Ibid., 38-41.
21. Ibid., 64.
22. Ibid.
23. Ibid., 64, 66.
24. Ibid., 67.
25. Ibid., 61-64.

4

The Physical Metaphor Illustrated

An Idea of Information-as-Thing: Shannon and Communication

The physical metaphor's propensity to reveal statistical regularities received a powerful stimulus from early research in communications engineering. Claude Shannon's work on mathematical communication theory represents a pure conception of information-as-thing, and contributed significantly to the development of the physical metaphor in information science. In fact, his model of communication is analogous to an information retrieval system, as illustrated by Figure 4.1.[1]

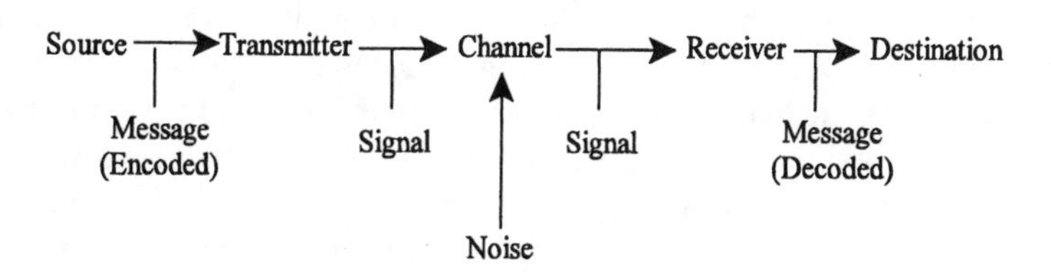

Figure 4.1 Shannon's Model of Communication

For the purpose of information retrieval, the source may be regarded as a collection of documents. Encoding the message becomes a matter of cataloging and indexing the documents; the transmitter, channel and receiver are analogous to the search engine; and noise represents the problem of excluding irrelevant documents. The signal to the left of the channel can be regarded as a query, and the signal to the right of the channel as a response. This system brings decoded messages in the form of documents to its destination, the user. Feedback, in this context, is first a matter of the user's judgment regarding the relevance of the information retrieved and, if necessary, it is followed by a reinterrogation of the collection by means of a new or revised query.

The central problem of communication, from the perspective of Shannon's model, is one of selecting a desired message from the set of possible messages. Questions must be answered about the amount of information that can be sent compared to what is actually sent, channel capacity, the coding process, and the effects of noise. Particularly crucial to the model itself is the unexpected and non-intuitive way the word *information* is used, in that meaning is entirely excluded. In Shannon's model, information is a quantitative measure of the freedom of choice available when selecting a message to be sent from the number of possible messages that could be sent. The concept that best describes this condition is entropy, i.e., the degree of randomness in a situation as expressed in terms of probabilities. When a situation is highly organized, and therefore not random, there is not a great deal of entropy, so the amount of information in the situation is low. In other words, the choices left to us regarding the number of possible ways the situation is structured are very limited.

As an illustration, let us say that I have eight marbles, each of which is encoded such that the eight marbles together constitute a message I want to send. I distribute each of the eight marbles at random throughout a room full of people, and they have thirty seconds to pass them around in such a way that all eight of them arrive in the hands of person in the back right hand corner of the room. Needless to say, the odds are low that all eight will arrive in time. Because there is a high amount of entropy in this situation, i.e., there are so many ways the marbles might be passed to achieve my goal, I would have to find a way to reduce the amount of information in the situation in order to ensure that enough marbles arrive with enough of the code that the person in the back of the room can interpret the message.

I am faced with three possible options. One, ensure that all eight marbles begin their journey at the same time and place, and that all are always passed together. This requires a great deal of control. Two, begin with a random distribution of the marbles, but double or triple the number of those that carry the same coding. This increases the redundancy of the system, and reduces its efficiency. Three, manipulate the characteristics of the marbles themselves to ensure that fewer of them are needed for the receiver of the message to make sense of it; this might require a considerable development of my coding system. All of these options have four things in common. They treat the message to be sent and the means of encoding it as *things*, in order to decrease the entropy of the communication system and increase its effectiveness. Together or singly, they impose a greater cost of some kind in order to accomplish communication. They all increase the probability that my message will get through, but none are certain.

Note that each member of the group charged with the responsibility of passing along the marbles need not even think about the meaning or content of the message. All they need to do is recognize a marble carrying a part of the message, distinguish it from any bogus marbles I may have introduced to confuse the situation (noise), and pass it along in the right direction. In a sense, they operate in the same way the devices and parts of an information retrieval system might work. It does not need to "understand" the query or the answer in order to properly handle them as objects of information. Any search engine operating on the principles of Boolean logic, for example, does not literally understand the meaning of the words in a search strategy or those assigned as index terms to documents. It simply seeks to match and combine them as given characteristics of queries and documents. Thus, the goal of a system designer, in the case our marble passing experiment, is to make it as easy as possible for each person in the group to recognize parts of the desired message and know what needs to be done with it even though no one but myself and the person in the back right hand corner of the room can really interpret its meaning.[2]

However, we are met with more ambiguity since users of information are rarely interested in the behavior of information in general. They are instead interested in the content of a specific message. They want to know and understand the specific messages encoded by the marbles, not where they've been. This is an issue to which we will return because it does not easily go away. But what if we were to design communication systems in which these people could be replaced by cybernetic mechanisms operated by software capable of recognizing the linguistically salient aspect of the message and determining what to do with it? The inclusion of automatic semantic mechanisms capable of interpreting the content carried by the marbles could increase the probability that a message selected by a destination is the one intended by a source, as well as the probability of desired information being retrieved. Probability then is the key to controlling information.

As it happens, language possesses a statistical structure compatible with this goal.[3] By exploring the relations of semantic arrangements and associations, it should be possible to arrive at conclusions regarding the relation between the amount of information and the amount of *meaning* in communication processes and, in so doing, reduce the amount of entropy and noise in information access and retrieval. Effective and efficient information retrieval systems, therefore, can be thought of as communication systems capable of sending and receiving small amounts of information (for example, a few search terms provided by user queries and index terms representing documents) in order to retrieve large amounts of meaning in the form of stored texts for users. There is an economy at work in this thought seeking to maximize the probability of retrieving the information desired by a user by reducing to a minimum the amount of entropy in the retrieval system.

In Shannon's model of communication, the ambiguity of "information" as a theoretical object can be eliminated by two moves. First, "information" is regarded as a general rather than a particular phenomenon to which neither text nor message is relevant in terms of content. It is a phenomenon naturally inherent in all communicative situations. Two, it can be operationally defined as a measure of entropy that itself characterizes communicative situations in general rather than as an attribute of any particular text or message. In this sense, "information" is a fundamental attribute of nature, like energy or matter. It is an inherent aspect and component of all physical reality.

A Method Based on Information-as-Thing: Bibliometrics

Alan Pritchard invented the term *bibliometrics* to describe "the application of mathematics and statistical methods to books and other media of communication."[4] The basic project of bibliometrics is to examine and analyze the artifacts of communication—typically written communication in scholarly disciplines—in order to arrive at conclusions regarding the nature of scholarly productivity and communication. Of central interest is the study of disciplinary development as signified by measures of production and use of scholarly literature.[5]

As a method of analysis, bibliometrics can be applied to any subject discipline. The theoretical objects appropriate to the method, including authors, titles, institutions, and citations, are self-contained, physical artifacts of communication, potentially informative and not independent of their meaning. But as objects of study and manipulation, they are separated from this meaning. This separation gives rise to the primary theoretical ambiguity of bibliometrics.

The research results produced as an outcome of this method are interesting and intriguing, but we might well ask what is actually revealed? Bibliometrics alone cannot say why some authors or journals are more productive than others or explain the observed ratios of productivity. Nor can it interpret productive regularities and citation patterns within scientific disciplines. To do so, the meaning of the observed artifacts must be returned to them by examining their context—the internal history of a discipline, the role of particular institutions and individuals, and so on. Still, bibliometrics has produced some fascinating results in two broad areas: bibliometric laws that describe the behavior of bibliographic objects, and citation analysis.

Bibliographic Laws

It turns out that books and other textual forms of communication display a number of general and quite regular behavioral characteristics. Books and journal articles representing scientific disciplines, for example, tend to distribute themselves in regular and predictable patterns. While the exact shape of these distributions vary by subject discipline, their patterns remain remarkably constant regardless of discipline. Robert Fairthorne identified the characteristic common to most of them as a "kind of hyperbolic distribution in which the product of fixed powers of the variables is constant. In its simplest discrete manifestation an input increasingly geometrically produces a yield increasing arithmetically."[6] The basic laws of bibliometrics were empirically derived by S. C. Bradford, A. J. Lotka and G. K. Zipf. These distributions or bibliometric "laws" can be described mathematically, but they are really quite straightforward.

Lotka's law of author productivity says that for any body of scientific literature, represented by articles in scientific journals, a substantial number of authors will contribute a single publication, a smaller number will contribute some articles, and a few will contribute a great number of articles.[7] Mathematically, Lotka's law is expressed this way:

Number of authors writing n articles $= 1/n^2$ (number of authors writing one article).

This law exemplifies what is known as an inverse square relation. If, for example, we have a body of scientific literature constituted by 1000 articles, and 600 authors have each contributed one article, then the number of authors contributing two articles ($n = 2$) will be:

$$1/n^2 = 1/2^2 = 1/4$$

and so,

$$1/4\ (600) = 150$$

One hundred fifty authors will have contributed two articles each. The number of authors contributing ten articles each will be six. Given that Lotka's law implies that any scientific discipline tends to be dominated by the work of a few people, knowing who those people are will be very important for collection development and information retrieval purposes.

However, Lotka's law is not without limitations. It tends to break down at the high end of author productivity, in that a small number of authors in most scientific disciplines tend to be much more productive than the law predicts. Also, despite the fact that Lotka's law makes sense intuitively, it is not at all clear why this distribution occurs. If author productivity is taken as a dependent variable, the jury is still out on what independent variable or variables explain the observed distribution curve.

Another early observation of the regular and predictable behavior of information is expressed by Bradford's law of scatter. This law is also intuitive and based on the observation that collection use is rarely evenly distributed. Generally a few documents will be used quite often, some documents will be used sometimes, and a great number of documents will hardly be used at all. Thus ,for any given subject and collection of relevant journals, Bradford's law states that it is possible to divide these journals into zones that, while containing the same number of articles on the subject, the number of journals needed to yield that number of articles increases substantially from one zone to the next.[8] Mathematically, Bradford's law is expressed this way:

If $R(n)$ = total # of journal articles,
N = total # of journals, and
s = a constant specific to a given subject discipline, then

$$R(n) = N \log n/s \, (1 \le n \le N)$$

Bradford's law may be used as a measure of the efficiency of journals yielding information relevant to subjects. A typical Bradford distribution might look something like Figure 4.2. For example, if someone needed information on this hypothetical subject, then the journals in zone one should be the first ones to be examined. True, the journals in zone three contain the same number of articles on the subject, but to examine all 250 titles might be prohibitively costly and possibly not worth the effort.

Zone	# of Journals	# of Articles
1	10	400
2	60	400
3	250	400

Figure 4.2 Bradford Distribution

Bradford's law is the basis of the well known 80/20 rule in librarianship which states that 80 percent of the use of a library collection can generally be accounted for by 20 percent of the collection. Another way of viewing Bradford's law is simply that most of the information relevant to any given topic will be found in a few sources, and to retrieve an equal amount of new relevant information after those sources have been examined will require the retrieval of a great many more sources than were retrieved the first time.

Zipf's law of the distribution of words in texts is based on the intuitive notion that people tend to use only a small part of the vocabulary available to them for communication.[9] For any text, it is possible to count the appearances of any word and to rank words by their frequency of appearance. Zipf's law says that for any given text, on any subject, if a rank order of the word's frequency is multiplied by the number of times it appears in the text, then the result is a constant. Mathematically, Zipf's law is expressed this way:

If f = frequency of appearance of a word, and
r = rank of the frequency of appearance of a word, then

$$(f)\,(r) = c$$

where c is a constant for a given text.

If, for example, in an article on reference work, the word *reference* appears more frequently than any other word, and it appears 66 times, its constant is:

$$c = 1(66) = 66,$$

which means that the next most frequently occurring word will appear 33 times:

$$2(33) = 66,$$

and the third most frequently occurring word will appear 22 times:

$$3(22) = 66.$$

If the second word is *librarian* and the third word is *user,* we might be tempted to infer that this article was *more* about reference work and librarians than it is about reference work and users, but users are not irrelevant to the topic. Clearly Zipf's law has implications for indexing and information retrieval, not the least of which is the possibility of establishing a quantitatively measurable probability of what a text is about and whether or not it would likely satisfy a user's need.

Underlying Zipf's law is economics of effort, specifically the idea that language is a phenomenon that seeks to convey the most information with the fewest number of words, and this insight delineates a puzzle to solve. Is it possible to identify a few words in any document that most likely represent the semantic embodiment of the document? Can this be done on the basis of Zipf's law?[10] Affirmative answers would have powerful implications for indexing and information retrieval. This puzzle represents a major focus of the physical metaphor, and its solution is the foundation for automatic indexing, but the inability to separate causes and consequences is a general theoretical problem for all of the bibliometric laws discussed here.[11]

Citation Analysis

Citation analysis is the study of a particular kind of sign within texts that indicates information has been used. The strength and the value of the results of citation analyses are due to the focus on objective, empirical, and quantitative evidence that can be statistically analyzed to yield generalizable knowledge and predictions about the structure of disciplinary discourses. The practice of citing prior work is an integral and fundamental aspect of scholarly communication, yet a citation can be treated as thing, and described in a language of physical things. The public nature of this practice is easily observable, and the study of citation behavior contributes to both information retrieval and to our understanding of the structure and development of research literatures.[12]

For example, a citation in one scientific paper of another clearly establishes a relationship as the connection between them allows us to infer a similar relationship in their content. This relationship may also clarify aspects of the aboutness of each document that subject descriptors leave ambiguous. It might even be possible to make a judgment about the value of the cited document to the author of the citing document. These relations, however, are the basis for citation indexing, which is a powerful means of information retrieval. Once we have examined the list of references an author has provided to support his or her text, we begin to understand where and how that author's text relates to the work of others in the same area of research. Such an examination will also help to establish the intertextual meaning of the text, as well as point to other information that the author believes to be important, useful, and pertinent.

Clearly, citation analysis is a powerful means of investigating the intellectual and paradigmatic structure of a scholarly discipline. It can reveal conflicts and alliances, agreed-upon and contested ideas, influences, schools of thought, and institutional arrangements. A text's citations, both from it and to it, locate that text in an intellectual space, and all of the data needed to identify this location is both public and can be gathered without the author's knowledge or permission. There is no need to penetrate the subjective world of the author nor make some inferences about private knowledge.[13]

From a practical point of view, citation indexing as a means of document retrieval avoids many of the troubling ambiguities associated with subject indexing, whether done automatically or by human beings. In effect, scholars index their own work by associating the subjects of the texts they cite with their own subjects, and they index the literature of their discipline by associating the subjects of their text with that of the texts they cite. In a citation index, the author's name substitutes for subject descriptors, representing content and a particular approach to content while avoiding the semantic and interpretive problems of subject indexing. As an added attraction, citation indexes can be created automatically and efficiently without knowledge of a discipline's subject.[14]

The logic of citation indexing is quite straightforward. Let us say that I have found an article that seems to meet my need. Still, as good as this article is, I feel a need to find more information of the same kind. A subject search based on keywords extracted from the article's title, abstract, or text is one possible course of action, or perhaps I have access to the index terms assigned to the article. On the other hand, I might simply ask myself, who has cited the article I have in hand? If someone has, it is likely to be the same subject and in some way would contribute to my thesis or idea.

This approach to document retrieval may be especially valuable if I know the article is useful to me but can't quite say why. It not only provides access to new material on similar subjects, it locates that material, and the original document, in an explicit intertextual context. While ordinary subject indexing can also provide intertextual context as well, it is implicit rather than explicit and requires considerable effort on my part become apparent. Once documents citing my original document are retrieved, I can repeat the citation search process with each of them. Conversely, I may explore the references in my original document and decide that some or all of them are of interest. Then, by checking them, I begin to build up a body of literature on a subject and possibly discern the outlines of particular, and perhaps conflicting, ways of thinking about it. This outcome is the explicit goal of citation analysis.

There are four fundamental methods pertinent to citation analysis. Simply counting the number of citations to a given item is most basic method. The cited item might be any work by a given author, a particular work by an author, a journal, or any other bibliographic entity. Although one must be careful about drawing oversimplified inferences, the number of times a text is cited is likely to be an indication of how important, influential, popular, or good it is. A closely related phenomenon is citation impact. This is a matter of counting the average number of citations to a given item in some specified length of time. This refinement is undertaken to introduce an element of control over intervening variables that can affect simple counting and bias. A journal, for example, may appear to be more influential than perhaps it really is because it publishes a great number of articles which produce a great number of citations to all those articles. Another journal may publish fewer articles, but those it publishes are truly influential and frequently cited. Nevertheless, it may not appear to be as influential as it really is because it publishes only a few articles and does not publish as often, and the absolute number of citations to articles that appear in it are less than the number of citations to articles that appear in the more frequently published journal.[15]

The remaining methods are bibliographic coupling and co-citation. The former is a matter of two or more documents being linked because they cite a third document. It assumes an intrinsic aboutness link between two documents that are based on the same prior work. Co-citation is a matter of two or more documents being linked because they are cited together by a later work. This condition assumes an extrinsic aboutness link between the documents that are cited together. Every one of these four phenomena can be measured and ranked quantitatively. At the same time, each of the phenomena is an indicator of the relevance of the documents to one another and potentially of documents to authors and readers. Each can be used to provide insight into the social and intellectual structure of subject disciplines.

The promise of citation indexing as a means of information retrieval and the promise of citation analysis to answer some very interesting questions regarding the nature of scholarly communication, are nevertheless diminished somewhat by some persistent operational problems related to the ambiguity of a citation as a theoretical object. Citation indexing, for example, must accommodate multiple authors. A citation index reference to only the first listed author overlooks the contribution of other authors to a text. Even when reference is made to each, it is impossible to identify their separate and distinct contributions relative to the text as a whole. Homographs (two different people who happen to share the same name) and synonyms (the same person whose name appears in variations) also cause problems of accurately linking cited and citing texts. In some discourses, implicit citations are prominent and often vital to their texts. Implicit citation occurs when an author assumes that his or her reader is a participant in the discourse and knowledgeable regarding it. The author then refers to an idea or another author but does not use a formal bibliographic citation. This kind of citation completely eludes citation analysis. Finally, the meaning of a citation can itself be ambiguous.[16]

There is no doubt that citation analysis can locate a given document in a literature that is constituted by documents about the same subject, but questions can be asked about the accuracy of that location and the nature and meaning of the relation between citing and cited documents. Not all citations are of equal value, and there are many reasons why an author may cite another text. The citation may be positive, negative, necessary, or pointless. It may be done to authenticate a claim, give credit where credit is due, levy a criticism, or serve less than noble reasons.[17] Especially when used as a means of judging the quality of the cited item, citation analysis requires careful interpretation.

Six assumptions underlie the performance of citation analysis, all of which are more or less related to problems of interpreting the meaning of a citation and in turn can be challenged. The first is concerned with how cited documents are used. Citation analysis must assume that a cited item was actually used by the citing author and contributed to his or her work. Given the wide variety or motivations of citing the work of others, this assumption is problematic to say the least. A work may be cited because it is expected of the authors. A work may be cited when actually only a very small part of it was actually relevant. Other texts may be used but not cited, or cited and not used, and this behavior may or may not be deliberate. A closely related assumption is that a cited text be related to the content of the citing text. Some studies of citation indexing have questioned the authors of cited documents regarding the relevance of documents that cite them to the subject of their original work. On occasion, those authors were unable to understand the reason for the citation and concluded that the citing document was not, in fact, relevant to the subject of the cited document. In the case of bibliographic coupling, two or more documents may cite a third, for example, but the reference may be to entirely different parts for entirely different reasons, distinctions not easily revealed by citation analysis.[18]

Two other assumptions are concerned with the value of cited documents. A document that is frequently cited is assumed to possess some quality that makes it seminal for the social and intellectual practice of a given discipline. A document might be cited, however, as an example of bad research or weak thinking. Judging the quality of cited texts simply from the number of times they are cited is a risky business. By the same token, we cannot assume that an infrequently cited text is without merit, since it is possible that its value has not yet been discovered. Even more problematic is the case in which a text is potentially of great value but not cited because it challenges the core elements of a dominant paradigm. Conversely, it is also possible that a text of little real value is cited often simply because it represents all that is acceptable to a dominant paradigm. Citation to this kind of text may in fact be necessary if a scholar is to be published in some journals.

A fifth assumption holds that citations should always be made to the most important and most useful texts available. In contrast to its underlying rationality, authors may cite texts because they are easily accessible or already known to them. This practice may also include self-citation. Thus, visibility or reputation of the cited authors has more to do with citations than their subject relevance. The motive for citation behavior may vary considerably , and the social context within which that behavior occurs can affect it in profound and not entirely explicit ways. Different disciplines and different cultures, for example, have different expectations regarding citation behavior.

A simple way of describing the grounds on which all of these assumptions can be challenged in turn challenges the sixth and final assumption; some citations are "more equal" than others. The question is: Which are truly relevant and meaningful to the citing text? The main theoretical flaw of citation analysis is that it produces an interesting but reified artifact. Social and intellectual structures can be clearly discerned, but explanations of the reason for and the meaning of those structures remain elusive and cannot be determined or inferred from the observed structure alone. Adequate and accurate judgement of the value of cited texts, is likely to require some method in addition to bibliometrics.

None of the bibliometric laws we have described, nor the admittedly intriguing findings of citation analysis, have come to serve as basic building blocks of an information science. Bibliometric laws and citation analysis are powerfully suggestive about where to look for explanations of the patterns they reveal, but they are silent regarding how to look for those explanations. There is some controversy regarding whether bibliometrics can actually serve as an explanatory theory or if it is a method in search of a theory.[19]

An interesting and generally dominant view of bibliometric phenomena is that they all tend to imply a theory of cumulative advantage similar to that of Pareto's Law of Income Distribution.[20] Derek de Solla Price explains the theory this way:

> A paper which has been cited many times is more likely to be cited again than one which has been little cited. An author of many papers is more likely to publish again than one who has been less prolific. A journal which has been frequently consulted for some purpose is more likely to be turned to again than one of previously infrequent use. Words become common or remain rare. A millionaire gets extra income faster than a beggar.[21]

With this statement De Solla Price confirms a common sense notion often expressed as the "Matthew effect," i.e., to those that have, more will be given.

On the other hand, this common sense notion, when applied to information, implies that information displays a regularity and predictability of behavior that can be described in a language of physical things.

Counting and statistically analyzing various aspects of written communication can contribute to our understanding of that behavior. The results of these studies may be descriptive rather than explanatory, but they can clearly be of use in our efforts to control information. Observations of the behavior of "information," however, beg interpretation, and their explanation is not entirely inherent within these observations themselves. Message and meaning, as well as ambiguities regarding the aboutness of the objects observed and their relevance to the observer raise questions that bibliometrics alone answer.

Both bibliometrics and citation analysis view behavior in terms of invariant and predictable attributes, but if we are to understand and explain this behavior we must on occasion focus our attention on specific aspects even as we ignore other aspects. And while this reduction probably does some damage to human reality, there are few alternatives. Still, these techniques reveal much about the nature of human communication. The code is complex, and there is no reason to assume that it will be easily penetrated; but like a painting or a piece of music, a pattern in the behavior of information represents a domain of thought that can tell us a great deal about the common reality we inhabit.

An Experiment based on Information-as-Thing: Cranfield

According to Ellis, the physical paradigm emerged from a reaction to conventional subject indexing and information retrieval established by library science and documentation. While these disciplines had developed extensive and logical systems of description, classification, and organization of information, there existed no empirical evidence of retrieval effectiveness of these systems. Other than conventional, expert opinion, there was no agreed-upon method of determining which system performed best. Discontent with this situation, combined with the new possibilities introduced by computer technology, led to the Cranfield experiments of the mid-1950s.[22] Directed by Cyril Cleverdon, these experiments aimed at establishing information retrieval research as an empirical discipline. They were based on the conviction that the relative merits of information retrieval systems must be judged on a scientific and empirical basis.

Specifically the Cranfield experiments studied the relationship between indexing languages for document retrieval and operational measures of their retrieval effectiveness.[23] Cleverdon and his team recognized that these efforts were an abstraction from reality, but they believed reduction to be necessary in order to clearly and objectively understand the relations between the experimental variables, as well as to unambiguously explain the results of the tests. The results, they hoped, would prove useful in the designing of actual operational document retrieval systems.

The ultimate goal of an information retrieval system is, at the very least, to provide access to texts that are relevant to the seeker's query. Thus, issues of representation and organization of texts, while of central importance, are only attempts at determining the best means to this end. The explicit purpose of the Cranfield experiments was to compare different means of representing texts in order to determine which of them performed best when retrieving texts in response to test queries. At the same time, the research team sought to establish an information science grounded in rigorous methods and empirical evidence.[24] To accomplish this, they needed to develop quantitative measures of information system retrieval performance.

In other words, they wanted to show not only that one method of text representation and retrieval outperformed another, but that it did so by a specified amount. Their hypothesis was that such measures would prove relatively straightforward and based on a seemingly simple question. How well does a particular system answer a question? It seemed reasonable to expect that experts in a given subject could come to an agreement about the relevance of documents to particular questions.[25] Documents whose index terms matched the terms presented by a query would either be about that query or they would not—a simple yes/no distinction. From this expectation came two basic measures of information system retrieval performance, each of which can be expressed by an apparently simple question. Of the number of documents in a collection relevant to a particular query, how many of them are retrieved? Of the number of documents in a collection irrelevant to a particular query, how many of them are excluded from retrieval? Both of these questions depend on the assumption that the relevance of a document to a query is not problematic. But we shall see there are reasons to believe that this assumption is on shaky ground.

One of the central methodological tasks of the Cranfield experiments was to ensure that the comparisons made among the means of representation under examination truly reflected differences in their ability to retrieve documents. Both the document collections and the queries asked of them had to be experimentally controlled in order to be certain that the results were due to the means of representation and were not related to the nature of the documents and queries themselves. To accomplish this task, Cleverdon's team selected highly specific test collections of documents selected for the experiments. The experiments were then designed to create a contaminate-free environment in which independent and dependent variables could be isolated and manipulated at will and from which subjective human judgment of results could be excluded.

In the first round of Cranfield experiments, the test queries were simply derived from the source documents in the test collection, in order to make certain whether it could be known in advance that the documents were about the queries. If the means of representation under investigation failed to retrieve a document in response to a query, it could be clearly determined if the cause was due to a failure of the means or to the fact that the document was truly not about the query. This approach to evaluation reflects a physical metaphor at work in the context of a controlled experiment.

In many ways Cranfield I, as the first round of Cranfield experiments have come to be called, was merely a test of experimental method. Nevertheless it became a focus of widespread commentary and criticism, specifically that the results of the experiments were merely an artifact of their design.[26] Of particular concern was that the intimate link between queries and source documents might have clouded the test results.[27]

Cranfield II was decidedly more sophisticated so as to address the criticisms of Cranfield I. Here, the test queries were not derived directly from the source documents. Instead the original authors of the test collection documents were asked to reconstruct the research questions that gave rise to the documents in the first place. To determine if a document was in fact about a query, all of the documents in the test collection that were judged relevant were collected and sent, along with the query, to the original authors. The authors of the documents, then were the final judges regarding relevance. Documents judged to be irrelevant were then discarded from the test collection. For any given test query then, the Cranfield researchers had already identified which documents in the test collection were relevant to a given query.[28]

Relevance-based quantitative measures of retrieval performance were defined and used to compare the effectiveness of different means of representation. These measures are recall and precision. Recall is a measure of the extent to which documents relevant to a search query are retrieved from a collection of documents. It is a measure of the ability of a system to retrieve relevant documents. Precision is a measure of the extent to which only documents relevant to a search query are retrieved from a collection of documents. It is a measure of the system's ability to exclude irrelevant documents from retrieval, not unlike a signal to noise ratio. These measures are ultimately based on a relation between terms used in a query and terms used to represent documents, and based on this relation, a retrieval system's ability to predict the relevance of a document to a query. This situation is illustrated in Figure 4.3.[29]

	User Relevance Decisions		
System Relevance Prediction	Relevant	Non-relevant	Total
Retrieved	*a* (hits)	*b* (noise)	$a + b$
Not retrieved	*c* (misses)	*d* (correctly rejected)	$c + d$
Total	$a + c$	$b + d$	$a + b + c + d$ (total collection)

Figure 4.3. Retrieval Results

If, in response to a query, a system retrieves a document that an end user judges to be relevant, then the document is a hit (*a*). If the user judges the document not to be relevant, then the document is, in effect, noise (*b*). If in response to a query, the system fails to retrieve a document that an end user would have judged to be relevant, then the document is a miss (*c*). If the system successfully avoids retrieving a document that a user would have judged to be irrelevant, then that document has been appropriately rejected by the system as irrelevant to the query (*d*). For a given search, $a + b$ represents all documents retrieved. In almost every instance, however, some of these documents will not be relevant to the query (*b*). For the same search, $c+d$ represents all the documents not retrieved, some of which however, will be relevant to the query (*c*).

Thus recall is mathematically defined as the number of documents retrieved divided by the number of relevant documents in the collection, or $a/(a + c)$. And precision is mathematically defined as the number of relevant documents retrieved divided by the total number of documents retrieved or $a/(a + b)$. Each of these fractions is then multiplied by 100 to arrive at a percentage. Recall is the percentage of relevant documents in the system that are actually retrieved in response to a query. Precision is the percentage of relevant documents among all of the documents retrieved. In a perfect search, *b* and *c* equal zero, i.e., no relevant documents are missed, and no irrelevant documents are retrieved. In this case, both recall and precision are equal to 100 percent. This outcome is rare. Of course, 100 percent recall can be guaranteed by retrieving every document in a collection, but this will result in very low precision. Precision, then, resembles a cost factor. How much effort is required of a user to examine all of the retrieved documents in response to a query to sort out which are relevant and which are not?

Precision is also a measure of the efficiency of a search. Given a fixed recall ratio, as precision decreases, so does efficiency.[30] For example, let's say we conduct three searches with the following results:

Search 1 retrieves 30 documents: Precision = 15/30 = .5 x100 = 50%

Search 2 retrieves 60 documents: Precision = 15/60 = .25 x 100 = 25%

Search 3 retrieves 150 documents: Precision = 15/150 = .1 x 100 = 10%

If we assume there are 20 items in a collection relevant to a query and each of three searches retrieves 15 of them, then the recall ratio for all three searches is 15/20 = .75 x100 = 75 percent. The lower precision of Searches 2 and 3 means a user of the documents will have to expend considerably more effort to identify and use the documents that are actually relevant to his or her query. Perhaps the most important conclusion, however, is the tendency for an inverse relationship between recall and precision.[31] The more a search is tailored to produce a high level of recall, the more likely it is that precision will be reduced. To focus a search in such a way as to maximize precision increases the likelihood that documents relevant to the query will be left in the system. The use of an *or* operator in a Boolean search of an online database, for example, broadens a search and increases the odds of improving recall at the cost of precision. The use of an *and* operator has the opposite effect. It narrows the search and so improves precision at the cost of recall.[32]

The development of these measures of information system retrieval performance have contributed significantly to our understanding of the relationship between representation and retrieval and have been invaluable to system design and testing. Still, there are a few caveats that ought to be kept in mind. Recall and precision should not be regarded as absolute measures of retrieval effectiveness. Different users can and do have different needs for recall and precision. In some instances, maximum recall may be less important to a user than retrieving two or three relevant items so as to gain immediate access to specific information. In others, the user may desire a comprehensive search and is willing to tolerate low levels of precision in return for being certain that little of importance is missed. Few people can specify in advance the specific levels of recall and precision they would find desirable. More likely, they can at best indicate the kind of results they think they need.

To paraphrase Orwell, while we have assumed that all relevant documents have equal value, we might more realistically assume that some documents are more relevant than others. In addition, the level of precision tolerance people display varies considerably, such that it cannot be assumed that search efficiency is a dominant value in every case.[33] Of course, outside of the environment of a test collection, it is practically impossible to know how many documents in a collection are relevant to a particular query. While sampling-based methods can be used to test the retrieval effectiveness of real information retrieval systems, they are difficult to apply in specific situations. Nevertheless, the logic of these measures can be a useful guide to devising a search strategy depending on the goal of the search; and one can structure a search statement based on its likelihood of generally maximizing recall or precision.

A more difficult problem to solve is posed by the implicit model of information-seeking behavior assumed by the Cranfield team's approach to evaluation. Following the physical metaphor, this approach necessarily engages in certain reductions of reality in order to be able to apply a rigorous and particular method to the study of information problems. Since the definitions of recall and precision are operational in nature, they are ultimately grounded on an unnecessarily narrow conception of relevance.

First, we must assume that users of information retrieval systems have explicitly recognized needs and come to information retrieval systems with specific queries based on those needs. Systems used will then match representations of queries with those of documents and retrieve the latter on the basis of similarity. After the system has presented its users with the documents most likely to satisfy their stated needs, these users must then examine them and make a yes or no judgment regarding their relevance. Second, we must also assume that users of information retrieval systems will make a series of discrete judgments regarding the relevance of each retrieved document and that their understanding of their needs will not change as each document is encountered. Otherwise, it is possible that their relevance criteria as well.

With these two assumptions in place, the validity of recall and precision as measures of retrieval effectiveness as well as the stability of relevance as a theoretical construct are ensured. The resulting model of information seeking is both logical and linear with a beginning, a middle, and an end. But as with Shannon's model of communication, there is little room for consideration of the ambiguities of "information," aboutness, or relevance that might be at play.[34]

This notion of relevance is referred to as the *system* view of relevance. It is based on the assumption that in a perfect information retrieval system, retrieved documents are relevant and those not retrieved are irrelevant. Relevance then is regarded a product of the system and its internal mechanisms, a simple topical match between a document and a query. Should this condition not be met, it is due either to the ineffectiveness of the system's means of representation or inadequacies in its application.[35]

Still, we should take care not to dismiss such ambiguities too quickly. If, for example, we accept the notion that information itself is not only a tangible object but also can effect change in the knowledge state of its user, then relevance becomes a sign of the manner and extent of this change. Rarely a simple yes or no judgment then, relevance comes in a variety of packages and is manifest in a variety of ways.

Carlos Caudra's notion of the "black box" of relevance will help to set this idea into context. Not long after the Cranfield tests, Caudra conducted some experiments that revealed that one could manipulate relevance judgments by altering the instructions given to judges, even if those judges were the original authors of the documents retrieved in response to a test query. By "simply telling judges how the documents are expected to be used," both query and document were interpreted differently.[36] Thus, in the "black box" things go in and things come out; and depending on the settings applied to the box, we can see patterns and relationships between inputs and outputs, yet what actually occurs inside the box remains a mystery.

To be fair, Cleverdon was not unaware of this problem. It is rare for any scientific experiment to adequately capture the nuances of the reality it seeks to explore on its first try, and the relevance judgment problem was a central consideration in the redesign of the Cranfield tests. The operational solution devised by the Cranfield team for assessing relevance may be seem overly simplistic, but it allowed them to create a quantitative, experimentally grounded measure of retrieval effectiveness.

Later research makes clear that any number of factors might be at play, including the nature and characteristics of documents and their users, thus exposing the limitations of topic-based system relevance.[37, 38] While an objective topical relation between a document and a query is an important consideration, the user's interpretation of this relation is equally important if not more so. Witness the birth of the *destination* view of relevance, which emphasizes the user's characteristics over the source of the information.[39]

Conclusion

We have now arrived at a point where we can see the essential weakness of the physical metaphor and the approach to the study of information that it suggests. The theoretical stability of the object that it posits for examination begins to break down under the weight of a plausible, different and unavoidable meaning of the signifier "information." This conclusion should not be taken to mean that the physical metaphor is in some way flawed to the extent that we cannot trust research on which it is based. The metaphoric and suggestive power of Shannon's model of communication, bibliometrics, and the Cranfield experiments all provide a solid foundation for the purpose of research into how and why information behaves the way it does. Shannon observed that communication systems are constituted by probabilities and entropy; bibliometrics found that scholarly communication can be characterized by mathematical formulations of the principles of cumulative advantage. The Cranfield experiments treated information retrieval systems as natural systems that could be reduced to basic elements and their relations. On the basis of assumptions provided by these lines of research, it is possible to study information with scientific methods appropriate to the study of any other natural object. The Cranfield experiments in particular have become an exemplar of information retrieval research, though not in the Kuhnian sense. And although bibliometrics cannot point to a single set of experiments as its foundational source, its methods have provided the basis for an ongoing research project. The application of these methods to a wide range of puzzles regarding the nature of scholarly productivity and communication as well as to problems related to understanding the nature of research disciplines and their development has come close to realizing many of the goals of Shera's social epistemology.

In essence, the physical metaphor brings a research focus to the logical relations between statements of originators of information and statements of requests for information as mediated through technologies of representation and retrieval of information. In general, the goal of research in information science based on these two broad ideas has been to create a knowledge of the behavior of information as a natural object in order to develop more effective and efficient technologies of its manipulation and control. In the process it does a number of things one might expect from a paradigm in the immature sciences: it attracts adherents, identifies puzzles to solve, generates work, and leads to experimental results which are then applied in operational information retrieval systems. Some information scientists, however, question whether the physical metaphor is reaching a point of diminishing returns, and that the future breakthroughs will require new and different ideas.[40] And while the behavior of information as a physical object can be more accurately predicted and adequately controlled as a result of work done within the context of the physical metaphor, theoretical explanations for its behavior remain elusive. The conceptual nature and structure of this metaphor disciplines research by limiting it to questions only about the output of information retrieval systems. By declaring subjective experience off limits for investigation, this metaphor prevents us from asking some very interesting questions about how that output is used.

It is clear that information is and must be associated with objects that exist independently of human consciousness, that informative objects possess properties that are independent of human perception, and that these properties can be objectively ascertained. It is equally clear that information can properly be regarded as a thing, but a thing that must be perceived and interpreted by a human subject before its informative potential can be realized. Someone must read a text, interpret a painting, or hear the notes of a song before the text, painting, or song has a meaning or use as information. The alternative to the physical metaphor's concept of information is based on the idea that all information is situational in nature, and the informativeness of any object depends on the circumstances of its context, perception, and use. "Information" is clearly a matter of the properties and qualities of objects that may be said to be informative, but it might also be a matter of the meaning attributed to those properties and qualities by human beings. While an informative object and its qualities may be constant and unchanging, the meaning of that object to a human interpreter is subject to all the contradictions and contests that might be brought to it by the context of interpretation. From this perspective, information-as-thing is a necessary but insufficient means of understanding "information" as a theoretical object. The physical metaphor's construction of "information" as a theoretical object is not inadequate, but it is incomplete.

Endnotes

1. Warren Weaver, "The Mathematics of Communication," *Scientific American* 181 (July 1949): 12-13.
2. I am indebted to Shawn Collins, Research Associate, School of Information Sciences, University of Tennessee, Knoxville for this analogy.
3. Weaver, "Mathematics," 15.
4. Alan Pritchard, "Statistical Bibliography or Bibliometrics?"*Journal of Documentation* 25 (December 1969): 348.
5. Alan Pritchard, "Computers, Statistical Bibliography and Abstracting Services," 1968; L. M. Raisig ,"Statistical Bibliography in Health Sciences," *Bulletin of the Medical Library Association* 50 (July 1962): 450, 461. Cited in Pritchard, "Statistical Bibliography," p. 349.
6. Robert Fairthorne, "Empirical Hyperbolic Distributions (Bradford-Zipf-Mandlebrot) for Bibliometric Description and Prediction," *Journal of Documentation* 25 (December 1969): 319-43.
7. A. J. Lotka, "The Frequency Distribution of Scientific Productivity," *Journal of the Washington Academy of Sciences* 16 (1926): 317-323.
8. S. C. Bradford, "Sources of Information on Specific Subjects," *Engineering* 137 (1934): 85-86.
9. George K. Zipf, *Human Behavior and the Principle of Least Effort* (Cambridge, Mass.: Addison-Wesley Press, 1949).
10. Donald B. Cleveland and Ana D. Cleveland, *Introduction to Indexing and Abstracting* (Littleton, Colo.: Libraries Unlimited, 1983), 149.
11. Danny P. Wallace, "Bibliometrics and Citation Analysis," in John N. Olsgaard, ed. *Principles and Applications of Information Science for Library Professionals* (Chicago: American Library Association, 1989), 19-20; and Daniel O. O'Connor and Henry Voos, "Empirical Laws, Theory Construction and Bibliometrics," *Library Trends* 30 (Summer 1981): 10-12.
12. Wallace, "Bibliometrics," p. 18.
13. Wallace, "Bibliometrics," p. 18-19; and Linda C. Smith, "Citation Analysis," *Library Trends* 30 (Summer 1981): 83-84.
14. Eugene Garfield, "Citation Indexes for Science," *Science* 122 (July 1955): 108-110.
15. Wallace, "Bibliometrics," 23-24.
16. Smith, "Citation Analysis," 86-90.
17. Ibid., 84.
18. Ibid., 98-100.
19. O'Connor and Voos, "Empirical Laws," 9.
20. Wallace, "Bibliometrics," p. 20.
21. Derek de Solla Price, "A General Theory of Bibliometric and Other Cumulative Advantage Processes," *Journal for the American Society of Information Science* 27 (September/October 1976): 292.

22. Karen Sparck Jones, "The Cranfield Tests," in *Information Retrieval Experiment*, ed. Jones , 256-283.
23. David Ellis, "The Physical and Cognitive Paradigms in Information Retrieval Research," *Journal of Documentation* 48 (March 1992): 51.
24. Cyril W. Cleverdon, "The Cranfield Tests of Index Language Devices," *Aslib Proceedings* 19 (June 1967): 173-194, and "Review of the Origins and Development of Research" *Aslib Proceedings* 22 (November 1970): 538-549.
25. David Ellis, "The Dilemma of Measurement in Information Retrieval Research," *Journal of the American Society for Information Science* 47 (January 1996): 25-28.
26. Jones, "Cranfield Tests," p. 271-272.
27. David Ellis, *New Horizons in Information Retrieval* (London: The Library Association, 1990), 1-6; and Jones, "Cranfield Tests," 268.
28. Ellis, *New Horizons*, 7-16.
29. F. W. Lancaster, *Information Retrieval Systems : Characteristics, Testing, and Evaluation* 2nd ed. (New York: Wiley, 1979), 113.
30. Ibid., 113.
31. Cyril W. Cleverdon, "On the Inverse Relationship of Recall and Precision," *Journal of Documentation* 28 (September 1972): 195-201.
32. Lancaster, *Information Retrieval*, 114.
33. Ibid., 117.
34. Ellis, *New Horizons*, 23-24.
35. Tefko Saracevic, "Relevance: A Review of and a Framework for the Thinking on the Notion in Information Science," *Journal of the American Society for Information Science* 26 (November/December 1975): 327.
37. Carlos A. Caudra and Robert V. Katter, "Opening the Black Box of Relevance," *Journal of Documentation* 23 (December 1967): 302.
37. Ellis, "Dilemma of Measurement," 27.
38. Taemin Kim Park, "Toward a Theory of User-Based Relevance: A Call for a New Paradigm of Inquiry," *Journal of the American Society for Information Science* 45 (April 1994): 136.
39. Saracevic, "Relevance," 328.
40. Peter Ingwersen, *Information Retrieval Interaction* (London: Taylor Graham, 1992), 61-63, 80-81.

5

The Cognitive Metaphor

A Critique of Information-as-Thing

The physical metaphor, and its emphasis on the nature of information-as-thing, is premised on the assumption that we can distinguish between tangible, formal objects, and expressions that possess the quality of being informative and intangible cognitive phenomena of knowing and being informed. The effort to separate communicative symbols from what they denote is not unlike trying to untangle a sign by conceptually separating the signified from its signifier. Michael Buckland, for example, asserts that despite the potential difficulties inherent in this effort, rules for drawing inferences from stored information can "operate upon and only upon information-as-thing."[1] Following this logic, knowledge can be represented, but only in the sense of information-as-thing. Knowledge representation, then, is of necessity both tangible and informative.

Buckland argues that the concept of "evidence" usefully identifies and describes the nature of information-as-thing. Evidence can exist in the form of documents, but it is not limited to that form. Events and natural phenomena can also serve as evidence in that they are physical objects whose characteristics and behavior provide testimony regarding the nature of reality. Information is like evidence in that it is passive—it exists independently of perception and need not be granted credibility or acceptance. Human beings do things to or with evidence. They examine it, interpret it, understand or fail to understand it, hide it, and sometimes even fake it.[2] Evidence has the capability of changing our knowledge or opinion but must be objective and tangible to be fully admissible. Consequently, a wide variety of objects can serve as evidence, and the concept of a potentially informative object need not be limited by associating it only with intentional communication.

Buckland also employs a very expansive definition of *document* which makes some information scientists uncomfortable. Any object that *documents*, that is, denotes, identifies, or signifies human knowledge or experience can be considered a document. Thus any physical information resource, artifact, model, or natural object can be a document, and its characteristics and behavior can be examined independently of its meaning. Questions of meaning are questions of "discourse" rather than documentation.[3] Accordingly, intentionality, need not be an aspect of an object for it to be informative—even if intentionally created, it need not be intentionally informative. An object may be regarded as informative so long as it is representative and serves as a representation. For example, a book is both an artifact and intentionally communicative; ships and buildings are artifacts, but not intended to constitute discourse; while animals are not artifacts at all, yet they possess representational power. All of these objects can serve as evidence, all are informative, and their status does not depend on whether anyone intended them to be informative nor perceived them as such. Buckland cites the "natural sign" as a long-established signifier of an object that is informative but without communicative intent.[4]

While a powerful and intuitively satisfying argument, there is an implication here that manifests an ambiguity. Is the quality that makes an object informative truly one that is inherent in that object, or must some subject first perceive and assign informative potential to an object? Should we view memory as information-as-thing? Memories can be collected, stored, and retrieved, and clearly they operate as representations, but does their subjective and cognitive status render them inadmissible as evidence? While Buckland's claim that objects can be informative without intending to inform recalls semiotics' contention that any object can and might serve as a sign, semiotics also emphasizes the role of an interpreting subject in the creation of a sign. An object acquires meaning, and so signifies and communicates by virtue of an interpretive intervention. Take Derrida's notions of writing. He asserts that writing is not merely a carrier of ideas. This could be true only if there is a constant one-to-one correspondence between word and thought, between signifier and signified.[5] Writing, and its meaning, may be divorced from the intentions of its writer but never from the intentions of its reader.

Such an argument challenges Buckland and raises doubt about limiting the theoretical nature of information to its status as a thing.

Can we be certain that a ship or building, let alone other objects such as hairstyles or clothing, is not intended to constitute discourse? Take a building, for example. To the contractor, the building may signify, at best, a project to be completed, a physical object to be constructed, and a source of income. The architect, however, is likely to be engaged in an explicit discourse with the city and the society in which the building will be located, and its design may be quite deliberatively communicative. Conversely, consider the possibility that an unfortunate soul in the grip of a psychotic episode discovers a piece of broken plastic in the street and, from this "evidence" concludes that he must abandon his desire for his brother's wife. In this case, an individual uses an object that has no meaning for anyone else to communicate to himself. Finally, consider an animal. The first time an animal is discovered and examined, it may have a tremendous informative effect and come to serve as a natural sign. But once the animal is known, and its properties and behavior documented, the nature and extent of its informative potential may be lessened.

What these examples suggest is that an object's quality of being informative, or potentially informative, is a condition of its relative autonomy. A potentially informative object may indeed be autonomous in the sense of physically existing apart from our perception or consciousness of its existence. On the other hand, its informative quality depends on the recognition of its informative nature. Until that moment, its status as a potentially informative object is irrelevant to its existence. Buckland confronts this possibility by conceding that "being evidence, being informative is a quality *attributed* to things."[6] The use of word *attributed* is telling. By whom? For what purpose? He knows that he must address the question of when an object is not informative. While we can always say that evidence ought to exist, or might exist, we are led to the unhelpful conclusion that "if anything is, or might be informative, then *everything* is, or might well be information."[7]

Buckland finally concludes "that the capability of 'being informative,' the essential characteristic of information-as-thing, must be situational."[8] The informativeness of any object is a matter of individual judgment and opinion regarding whether the object is pertinent, significant, and important to the one who perceives it. However, the informative object, like the sign, is a creation of interpretive consciousness, representing a dialectical synthesis of object and subject bound together by a code that allows its interpretation. Without consensus concerning what is informative, we would be unable to "progress beyond an anarchy of individual opinions concerning what is or is not reasonably treated as information."[9] Information, as both a theoretical and a practical object, must be in some way perceived and interpreted by a human subject before its informative potential can be realized. Another way of describing the categorical duality of information as a theoretical object is to say that information must be a thing, but that is never all it can be.

Information and the Cognitive Metaphor

The cognitive metaphor begins with the assumption that information is a situational theoretical object. From this perspective, the concept of information-as-thing is too restrictive to allow for the investigation of issues related to the use and effect of information on users. Based on an entirely different set of assumptions about the nature of information, the cognitive metaphor is nevertheless still concerned with the representation and their retrieval, but does not regard the physicality of text as an especially interesting phenomenon.

As noted earlier, a variety of objects and expressions, including documents, are potentially informative, which is what allows them to be characterized at texts. According to the assumptions of the cognitive metaphor, texts possess the potential to inform because they represent cognitive structures. A document, a painting, or piece of music can be thought of as texts because they represent ideas, thoughts, and knowledge deliberately and intentionally crafted into particular arrangements of meaningful signs by their creators. In this form they are capable of being used to effect a change in the cognitive structure of one who reads a document, views a painting, or listens to music. From this perspective, the problem of information is not merely one of effectively representing and retrieving texts, but understanding the discovery, the communicative nature, and interpretation of texts that can accomplish cognitive changes desired by seekers of information.

The meaning of *information* is central to the cognitive metaphor, as was not the case, for example, with Shannon's model of communication. Information is viewed *as* meaning where meaning involves the active attribution of informative potential to and interpretation of the object's informative structure. Potentially informative objects, not all of which need be texts, are message sources. Messages might come from the physical world in the form of empirical observations of characteristics and behavior of natural objects, or they might come from texts in the form of human communication.

The idea of the natural sign is not foreign to the cognitive metaphor, although its status as a sign and its meaning must be derived by an interpreting subject. A sign, whether naturally or intentionally communicative, must be "read," interpreted, and translated into meaning before it can be comprehended and understood as information, which is accomplished by engaging the influence of a "reader's" situation. The nature of the object itself, the circumstances surrounding the reader's encounter with the object, and the reader's affective, cognitive, and social state at the moment of encounter can all ~~can~~ affect how the reader interprets and constructs the meaning of the object. This condition describes what it means for information to be situational. However, since it is unlikely that we all share exactly the same situation, it is quite possible that each of us will interpret a given informative object in a different way. In other words, the same informative structure will likely have a different meaning for each of us and may actually represent different "information" for each of us. If this is true, then "information," as a theoretical object, cannot be understood solely in the language of physical things.

For instance, upon looking at a given book, we may all agree that it is the Bible, but from that point on its informative potential and meaning may have more to do with our beliefs that its characteristics. Cognitively speaking, information must be viewed as a form of knowledge which informs by reducing uncertainty, solving problems, or otherwise serving to reveal to us the nature of our reality and contributes to our understanding of our world and ourselves. In this sense, information is a matter of effect and depends on how we use it to attribute meaning to the world and our lives. It is quite literally the means by which we change our minds by reinforcing and challenging the mental and conceptual categories that form our world views. The development and restructuring of this model as we move through life occurs whenever we send, receive, or employ communicative messages. By this process we transform information into knowledge.[10]

A central assumption of the cognitive metaphor is that information processing and communication are necessarily mediated by systems of conceptual categories which for the processing device (human or machine) model the world that it inhabits. This model must include concepts of the process, content, and context of communication.[11] In other words, communication and its informative content are mediated by the information user's cognitive model or "image structure"of the world and conception of his or her own place in that model.

While this notion tends to blur some interesting possible distinctions between machines and human beings, it is a powerful representation of how we structure facts, perceptions, and values related to our lived experience. We become informed when communication results in changes in these structures, just as existing structures can have a significant effect on how information is used, and on the nature of the informing that occurs. For example, not all information that is communicated is received, or if received, is always treated equally. Information can be noticed or not noticed, accepted or rejected, but it must be assessed, characterized, and classified in the process of communication.

Selective perception, cognitive dissonance, and the condition of being open or closed-minded are examples of phenomena dependent on one's cognitive model of the world, which in turn depends on one's knowledge, beliefs, values, and the relations among them. Thus, the successful retrieval of texts is a crucial but partial step in the effective transfer of information since it does not address how and why those texts affect their users. To be intelligible, let alone useful, information must first be structured and organized for use by systems of mental categories and concepts which themselves are products of prior communication, information, and cognition.[12] Understanding how and why these structures influence information seeking and use is central to the cognitive metaphor of "information."

The key to understanding this metaphor's effectiveness resides in information's potential power to transform the cognitive structures of its users. On the one hand, it encompasses the deliberate and purposeful structuring of messages that are intended by their senders; on the other, the deliberate search for such messages. In either case, it presupposes that the person who structures and sends a message has some knowledge of the purposes of its intended receiver, and that the receiver is aware of both the sender and his or her purpose. From the cognitive perspective, then, information and its effect cannot be separated, and its effect depends on the context within which it is retrieved and used.[13]

The diverse practitioners who use the cognitive metaphor agree that information science should be about communication, and that its focus should be on the act of becoming informed rather than on information itself as an object. The emergence of the cognitive metaphor was a reaction to the limits set on the kind of questions that could be addressed by the physical metaphor, and it serves to organize a great deal of research. Its focus on users and the use of information, knowledge, retrieval heuristics, and human-computer interaction is both compelling and powerful and it leads to the study of a number of real problems not posed by research within the context of the physical metaphor.[14]

Perhaps the primary difference between the physical and cognitive metaphors is that the former focuses on the source of communication while the latter focuses on its destination. While source and destination are crucial components for both metaphors, the connection between them remains implicit in the physical metaphor. The retrieval of a relevant document represents an effective communication between a retrieval system and its user and, ostensibly, between the author of the document and its reader. In contrast, the latter outcome is not taken for granted by the cognitive metaphor. Instead, the connection between the source of information and its destination is explicit and the central research focus. The distinction between the two metaphors reflects the categorical duality of information.

Nicholas Belkin, for example, insists that information retrieval begins with a need for information rather than a query. When individuals pose their queries to information retrieval systems and seek out texts, they are initiating communication with the system—perhaps looking for a solution to some personal problem. Information scientists working under the guidance of the cognitive metaphor of "information" seek to design their retrieval system so that it informs as well as retrieves.[15] They are less interested in how well a system retrieves than in how well it informs. Matching terms in the statements of originators with the queries of requesters of information is, at best, a stand-in for the actual purpose of information retrieval systems which is to match the image structures of texts and users of texts. Thus improvements in information retrieval will come not merely from more sophisticated retrieval algorithms and indexing methods, but also from a better understanding of how and why human beings need and use information.

Finding a way to build a model of a user's cognitive state, however, continues to be a major stumbling block for research and system design. Efforts to date include the use of human-machine dialog, linguistic analysis of user need statements, or some combination of both methods to construct an image of the user's need for comparison to term and subject associations in texts. The goal is to match a model of the user's need with models of the knowledge inherent in texts.[16] While research based on the cognitive metaphor displays a thematic and theoretical consistency, the diversity of its objectives and methods has made generalization next to impossible. The key to understanding these problems lies in the inherent complexity and ambiguity of the central concepts of the cognitive metaphor. Concepts such as "image structures" and "becoming informed" may reflect the complexity of human information reality, but the cost has been a conflation of variables that make it difficult to isolate and sort cause from effect. One result is a dearth of practical applications.

Nevertheless, if the purpose of an information retrieval system is not merely to retrieve documents in response to requests, but to communicate information to its user, then texts, their use, and their users must ultimately be understood in the context of communication. This condition implies a need to explore the nature and intentions of the creators of texts as well.[17] In contrast to the physical metaphor's reduction of information to a material and manageable unit, the cognitive metaphor more closely reflects the complexity inherent in an intuitive understanding of information.

From the perspective of the cognitive metaphor, information is manifest in the cognitive structure of a text and capable of changing the cognitive structure of a reader. A text is a collection of signs purposefully structured with the intention of causing such a change. Phenomena of interest to information science from this point of view include but are not limited to the nature of texts as knowledge structures, the representation of knowledge structures, the cognitive structure of users of texts and how these structures are changed by use of texts, and the movement of texts between their producers and their users as part of a social process of communication. Meaning, interpretation, learning, the nature of knowledge and knowing, the effects of using information, and a host of other subjective phenomena are of paramount importance and can be represented by a simple yet powerfully suggestive fundamental equation that perhaps conceals more than it reveals.

The Fundamental Equation

The general focus of the cognitive metaphor of "information" is on human communication systems in which texts play a central role. In the mid-1970s, B. C. Brookes offered what he called "the fundamental equation" in order to describe the relation of knowledge and information and to define information retrieval in the context of human communication.[18] His fundamental equation is:

$$K[S] + \Delta I = K[S+\Delta S]$$

A knowledge structure, $K[S]$ is transformed into a new knowledge structure $K[S+\Delta S]$ by the addition of information, ΔI. Information, then, is a change agent that affects the content and organization of human knowledge structures and is, as such, a cognitive phenomena. Nor is knowledge scattered about the brain but is cognitively organized.

Information is thus itself a cognitive phenomenon, and it has the power to change what one knows, what one believes, and even what one values, as well as the relations among these mental phenomena.

Brookes makes three important concessions regarding the fundamental equation. First, it is not meant to be taken literally. It is a metaphorical expression that draws attention to what we must ponder in order to understand the nature of "information" and its effects on its users. Second, the interpretation of the empirical meaning of this equation, and the identification of its implications for information retrieval, must become the central research task. This metaphor suggests puzzles to be solved, and it symbolizes the problems to be investigated. Nevertheless, it requires that information be regarded in the same light as knowledge. In other words, if knowledge and information can be regarded as measurable objects, the same unit of measure must apply to each. Finally, there is no reason to believe—nor should this equation be taken to imply—that a given unit of information necessarily leads to a given change in a knowledge structure. Even within the same culture, a given unit of information may lead to different changes in a knowledge structure, depending on the knowledge structure under consideration, and the context within which it was received.

Nevertheless, this abstract formula leads to some very concrete and practical questions for research in information science. How do we assess the ways and extent to which a knowledge structure, a $K[S]$, is incomplete? How do we assess and identify someone's information needs? How do we identify, access, and evaluate ΔI, the information that will address those needs? How do we determine if ΔS, or the change in someone's knowledge structure, is adequate? How can we be certain that an information need has been satisfied? Remember, the central importance of the fundamental equation is that it reminds us that people do not *need* texts. They need the information structured by and contained in texts. From the perspective of the cognitive metaphor, document retrieval is only a means to an end, not an end in itself. Texts that are relevant to the topic of request may not be the texts that address the actual need for information, for the information they contain must be assessed in the context of a reader's current knowledge structure. The cognitive mediators comprising that structure, and through which a reader perceives and interprets a text, may or may not allow its contents to be transformed into information for that reader. In other words, successful information retrieval is a function not only of the objective characteristics of the information, but also the subjective characteristics of the relation between texts and their readers.

To better grasp this idea, it is useful to know that Brookes based his thinking on Karl Popper's theory of knowledge.[19] Brookes claims that the problem of information is fundamentally a problem of knowledge, which in turn is a problem of the nature of reality. According to Popper, real existence for human beings depends on three constitutive realities, or Worlds of Knowledge: what is, what we know of what is, and the formal expression of our knowledge. Together, these "worlds" constitute *the* world of human reality. World One, the world of what is, is the physical world around us and consists of all that may be known. It is vast and, despite our best efforts so far, much of it—likely most of it—remains unknown. It is a world of potential knowledge that is explored and revealed by means of scientific research.

World Two is a world of subjective human knowledge, a private world of the mind—in effect an individual world that exists within and only for each individual human being. It is the world of cognitive structures, knowledge structures, mental images, and mental states, populated with ideas, concepts, notions, perceptions, beliefs, values, frameworks, the mental apparatus and contents of cognition, and in some instances describable only by metaphor. Existing prior to formal or linguistic expression, this world is the one that each of us personally "knows." Finally, World Three is the world of objective knowledge, a public world consisting of objects that are products of the human mind. These objects include the physical metaphor's informative objects of texts, documents, and documentation. They are formal expressions of knowledge represented by tangible artifacts, and recorded in language, mathematics, statistics, art, music, and technology.

Each of these "Worlds" interacts with the others yet is also autonomous. The nature of this relation between Worlds One and Two is revealed in the classic categorical duality of body and mind. Each depends on the other but neither is merely a manifestation of the other. The body is of World One, the mind of World Two. The former is the location of sensory experience, the latter of the interpretation of that experience. For each of us, our reality depends on their differences and their interaction. The case for World Three is less obvious. A good analogy is this book. As long as I was writing it, it was mine to do with as I pleased—to revise, reject, and rethink. Once published, however, my book is on its own. It now belongs to its readers, and it can and will be read and interpreted in ways beyond my control or intention.

Worlds One and Two, and their relations, provide the focus for human research and the work of the arts, sciences, and humanities—the record of which is then deposited into World Three. The practical work of librarianship is to collect and organize the texts in World Three in order to enable their use. The theoretical and research work of information science is to study the relations and interactions between Worlds Two and Three in order to optimize the organization and use of "knowledge."[20] Thus the goal is to move beyond the bibliographic control of texts and gain control of the knowledge humanity creates and represents in its texts.

However, the problems in doing so are twofold. First it requires that information science develop a means to observe World Two, a problem which the physical paradigm maintains is insoluble at this time. Texts, most commonly but not necessarily in the form of documents, are objective events in World Three that represent the subjective constructions of World Two, but they do not provide a direct access to World Two, and they have problems of their own. World Three is not a tidy, complete, and discrete world. Its artifacts are often contradictory, self-challenging, and self-referential. They are not free of the influence of the purposes of communication. They do not necessarily represent "knowledge" in the way that word is conventionally conceived, and they may in fact express deliberate lies whose purpose is to deceive or oppress. The material existence of texts as objects in World Three does not imply that their contents are either objective or true. As a result, World Three presents significant challenges to both librarians and information scientists. Unbiased selection and representation for the purpose of building collections and providing access to texts may be very difficult to accomplish.

Nevertheless, the fundamental equation is not offered as a solution to the problems of the relations between Worlds Two and Three. Rather, it is a guidepost that suggests where to look for possible solutions. $K[S]$ and $K[S+\Delta S]$ are knowledge structures that exist in World Two. ΔI is the information represented in records and artifacts that exist in World Three. The fundamental equation directs attention to the relationship between Worlds Two and Three, and to the purpose of communicating information. From this perspective, the physical metaphor's concentration on the physical manifestation of "information," ΔI, is necessary but not sufficient to understand the problem with which information science really must be concerned—the relation between information and human purposes.

This approach recasts the problems of information science as matters of facilitating effective communication rather than as merely the effective and efficient organization and retrieval of texts. Information science must then study and find ways to facilitate the transformation of knowledge structures, $K[S]$ to $K[S+\Delta S]$, by means of the communication of desired information, ΔI, between a creator and user of knowledge.[21] The solution of this puzzle requires a very different concept of information than that posited by the physical metaphor.

To begin with, we must understand that while World Three is not a given phenomenon of nature, neither is it accidental in either content or structure. The objective knowledge of World Three, and the forms which contain it, are deliberately created and sought after in order to cause changes in World Two. From this perspective, information must be conceptualized in terms of purposeful and meaningful social communication. In other words, we must begin with the assumption that information is something deliberately produced by its creators from their knowledge structures with the intention of interacting with the knowledge structures of its users. Once knowledge is in an objective form, it does not fall automatically into the knowledge structures of various users. To retrieve knowledge from World Three in order to make it a part of his or her World Two, an information user must take a positive action. However, to complete the communication process, a potential user must deliberately access World Three and creatively interpret the contents of the texts he or she finds there. Information, then, is something desired, deliberately sought after, and acted upon by unique users in unique contexts. This move accounts for the nature of information, I, as ΔI, deliberately provided and sought after to accomplish certain desired changes.

The fundamental equation, however, also highlights the relationship between information as a change agent and the changes that result in knowledge structures by its use, that is between ΔI and ΔS. As we must consider how World Three, when accessed, changes an individual's World Two, so must information science account for the relations between information and states of knowledge, between I and $K[S]$, including the relationship between the creator and user of information. It must account for the effect of information on its user and the different effects it may have on different users in different contexts. Note that this idea draws attention to the right side of the fundamental equation, $K[S+\Delta S]$, and the effect of World Three, information, on World Two, knowledge.

While the concept of information discussed to this point relates it to the problems of effective communication, in order for it to be useful for practical applications and technologies, it must meet two operational requirements. It must be generalizable beyond individual cases; and it must allow us to predict the effects of information in general. The concept of information inherent in the physical metaphor meets these two operational requirements, but does not meet their theoretical counterparts. The physical metaphor directs our attention exclusively to World Three, treating information as an invariant characteristic. It does not distinguish between the content and cognitive structure of the information in texts, and because it focuses on retrieval rather than communication, it neither raises meaningful questions nor poses interesting puzzles about the nature of the relations between Worlds Two and Three.

In contrast, the cognitive metaphor emphasizes that information is more than merely an material object, and that retrieval is as an intentional action initiated by a user who is motivated by knowledge rather than texts.[22] Under its influence information science is focused upon the intentions of information seekers and the reasons for their behavior. What is it about a World Two that leads it to seek a relation with World Three? Why do people seek information and what do they expect to gain from its retrieval and use? What kinds of effects are caused by what kinds of information, and are these the effects desired by information seekers? By way of an answer, Belkin proposes what he calls an anomalous state of knowledge (ASK).[23] This concept presumes that the primary motive for seeking information is to resolve an anomaly in the seeker's cognitive structure, and that acquiring information does this by causing a change in content and/or structure of the seeker's state of knowledge.

If, for example, my cognitive structure and the model of the world that it sustains for me is challenged by an encounter with the unknown, I will likely experience some uncertainty. Perhaps it is caused by my inability to cope with the unknown, perhaps by an encounter with an informative object that I know I must attend to. Either way, I don't know what to do with the situation that confronts me. It doesn't seem to belong to any of the current categories or concepts that I ordinarily use to organize my reality. We might say that my current state of knowledge, my *K*[*S*], becomes damaged and in need of repair. I may be able to deny this condition for awhile, or delay the repair work, but unless the problem somehow takes care of itself, which is unlikely, this damage to my knowledge structure will soon cause me some trouble.

This trouble, then, represents the anomaly in my state of knowledge. I may experience such an anomaly as a gap in my knowledge, as a confusion about facts, a challenge to my beliefs, or a subversion of my values. I may experience it as a contradiction or an inability to choose between alternatives, and it may be an affective experience as much if not more than it is an intellectual one. If we conceptualize information as a product from World Three, capable of effecting change in the subjective knowledge structure of my World Two, then I can use it to reveal and resolve my ASK. This change may not be what I expect. By acquiring new facts, I may change my beliefs, but in any case, I will have changed my mind at least to the extent of resolving the anomaly that motivated my search for information.

The cognitive metaphor challenges a number of traditional assumptions in information retrieval research. Specifically, it challenges the idea that the relationship between a request for information and the need that motivates that a request is straightforward. An anomalous state of knowledge will not necessarily immediately lead to a determination of a need. It may not be easy or even possible for individuals to specify their needs, especially in terms of the options presented to them by information retrieval systems. Information needs may not be functionally equivalent to texts. Thus, in shifting the theoretical focus from World Three to World Two, from texts to users, from source to destination, the cognitive metaphor also shifts the focus of practice from representing texts and matching them with representations of queries to representing the ASK of each user and mapping them onto the knowledge structures of texts.

Conclusion

The strength of the cognitive metaphor is that it shifts our attention from the aboutness of documents to the structure of texts, re-establishing the intimate link between the need for information, the situation that gives rise to that need, and the way texts are used to address it. Doubt, uncertainty, and suspicions of inadequacy are seen not as noise, but as the very *reason* for using information retrieval systems. The cognitive metaphor suggests that the purpose of information retrieval is to help solve problems rather than to merely find texts about those problems. The user and the information retrieval system then become partners in a mutually adaptive dialog that allows and facilitates the creation of new knowledge structures. Whether the system involves artificial intelligence or a human librarian, it should not sit waiting for a request, but be available to intervene as soon as an ASK begins to form. The implications for institutional practice are profound and point to a redesign of services that would involve information services and sources to an even greater extent in our everyday lives.

However, the cognitive metaphor does not yet have an answer to the question of how this redesign is to be accomplished. While it provides a rich ground for theoretical speculation, it has yet to lead to the development of workable systems whose merits are clearly superior to the information retrieval systems that have been built based on research conducted in the context of the physical metaphor. There are theoretical difficulties as well. Both the fundamental equation and Popper's Worlds lack evidence of their physical existence. These concepts help make sense of experience, but use of the signifier "structure" to talk about knowledge imputes a concreteness to "knowledge" that it may not possess or deserve. As a theoretical object then, "knowledge" manifests the same if not greater ambiguities than "information," and if, as the fundamental equation implies, knowledge and information are phenomena of the same kind, and measurable by the same unit, what really is their difference? In recent discourse the term *knowledge retrieval* has appeared, and while this metaphor also has an intrinsic appeal, can it truly be "retrieved," or must it be earned, worked for, and tested?

Nevertheless, the fundamental equation poses some very interesting questions for research. Does retrieved information lead to desired and needed changes in a user's cognitive structure? Does it actually help a user solve a problem? What is the relationship between particular cognitive changes and the content, methods, approaches, and styles of formal expressions of knowledge? Why do these particular changes occur and not others? Why is one person affected in one way, and another in a different way, when exposed to the same information? Can the essential idea behind the fundamental equation be extended to include the notion of transforming the collective knowledge structure of society?

These questions hold more than a philosophical interest. Their answers can have immediate implications for the evaluation of information retrieval systems. An ability to measure the effect of information on cognitive structures would be of great value in information retrieval situations. This kind of system feedback could improve information retrieval in specific searches or in general, by finding information that actually addresses needs rather than merely fulfills requests. The more we know of the cognitive structures of users, their knowledge and the gaps in that knowledge, and the reasons why they seek information, the more effective will be our retrieval efforts.

The next chapter takes a look at a theoretical extension of the cognitive metaphor and two examples of research inspired by it. They address three central aspects of "information" as signified by the cognitive metaphor: the need for information, the information seeking process as motivated by that need, and the belief that information use can help one to make sense of the world. All three of these examples are grounded on the assumption that is, in fact, possible to empirically assess, if not directly observe, an individual's cognitive state of knowledge. All depend on a methodology of observation and inference that would not satisfy an information scientist seeking to develop an experimental science. However, each, in its own way, explores alternative significations of "information." They are linked not by the act of retrieving informative objects, but by what occurs and what must occur prior to and during that act.

Endnotes

1. Michael K. Buckland, "Information as Thing," *Journal of the American Society for Information Science* 42 (June 1991): 351-352.
2. Ibid., 353.
3. Ibid., 354.
4. Ibid., 355.
5. Jacques Derrida, *Deconstruction in a Nutshell: A Conversation with Jacques Derrida*, ed. with commentary by John D. Caputo (New York: Fordham University Press, 1997).
6. Buckland, "Information as Thing," 356. (Buckland's emphasis)
7. Ibid., 356. (Buckland's emphasis)
8. Ibid., 356-357.
9. Ibid., 357.
10. Gernot Wersig and Ulrich Neveling, "The Phenomena of Interest to Information Science," *The Information Scientist* 9 (December 1975): 130-132.
11. Nicholas. J. Belkin and Stephen. E. Robertson, "Information Science and the Phenomenon of Information," *Journal of the American Society for Information Science* 27 (July/August 1976): 201-202; and Nicholas J. Belkin, "The Cognitive Viewpoint in Information Science," *Journal of Information Science* 16 (1990): 11-12.
12. M. DeMey, "The Cognitive Viewpoint: Its Development and Its Scope," in *International Workshop on the Cognitive Viewpoint* (Ghent: University of Ghent, 1977), xvi-xxxii.
13. Belkin and Robertson, "Information Science," 197-201.
14. David Ellis, "The Physical and Cognitive Paradigms in Information Retrieval Research," *Journal of Documentation* 48 no. 1 (1992): 53-55.
15. N. J. Belkin, R. N. Oddy, and H. M. Brooks, "ASK for Information Retrieval: Part 1. Background and Theory," *Journal of Documentation* 38 (June 1982): 62-66.
16. N. J. Belkin, H. M. Brooks, and P. J. Daniels, "Knowledge Elicitation Using Discourse Analysis," *International Journal of Man-Machine Studies* 27 (1987):127-129.
17. Belkin and Robertson, "Information Science," 201-202.
18. B. C. Brookes, "The Foundations of Information Science. Part 1. Philosophical Aspects," *Journal of Information Science* 2 (October 1980): 131.

19. Ibid., 126-128; and Karl Popper, *Objective Knowledge; An Evolutionary Approach* (Oxford: Clarendon Press, 1972).
20. Brookes, "Foundations," 128, 132-133.
21. N. J. Belkin, "Information Concepts for Information Science," *Journal of Documentation* 34 (March 1978): 56-59.
22. Ibid., 80-81.
23. Ibid., 81.

6

The Cognitive Metaphor Illustrated

Negotiating Needs

Whether the issue is one of research or practice, the cognitive metaphor suggests that a holistic view of the user of information is required. A significant step in this direction was taken by Robert Taylor in his study of information-seeking in libraries. According to Taylor, when information seekers use a library, or any formal information system for that matter, they enter a negotiation, first with the system itself, and then with the authors of texts. Often, in order to access these texts, users of libraries must engage in a communication process with librarians that Taylor calls question negotiation, during which a librarian works with an information seeker to formulate an answerable question.[1]

One way of thinking about this situation is to imagine someone living within the context of what can be called a problem space. This space is typically configured by the nature of work an information seeker wants or needs to accomplish.[2] On a college campus, for example, this work will likely be a matter of competing assignments, writing papers, and doing formal research. The kinds of assignments and research goals set for people and the requirements and expectations that must be met constitute the problem space or, in other words, the source from information needs arise.[3] In order to actually look for information, a query space must be created from this problem space. The query space is configured by the specific need arising from the problem space, its informational aspects, and the nature and form of the information that likely will satisfy the need. Once this space is created, and an answerable question is asked, the search for information can begin.

This metaphor suggests that a search for information is not and should not be thought of as a single discrete event. Information seekers experience doubts which are open-ended, dynamic, and negotiable. Likewise, an anomalous state is not a specific and unchanging characteristic of the information seeker. Thus, creating a query from an information seeker's problem requires an open-ended and dynamic process of negotiation between the seeker and, often through a representative, an information system.

Let us return to our first example. Being aware that a library is a potentially useful source of information represents a significant step in the development of a query from a problem, just as reaching a librarian to begin a question negotiation process represents the outcome of a relatively complex sense-making process. Note that both of these outcomes occur before the information seeker has asked a librarian a single reference question. Still, they indicate that the information seeker has begun to construct a problem space that includes a need for information. The first step toward making sense of a situation or solving a problem that does not make sense is to know what needs to be done or asked. An anomaly can be rendered less so almost as soon as it is understood to *be* an anomaly.[4]

Just as information search is not accurately conceptualized as a single event, neither is an information need a single instance. Taylor found information needs to exist at many levels, not all of them in support of an answerable question.[5] In its earliest articulation, an information need is *visceral*. For example, my experience of a need at this level is more a matter of feeling than thought, manifesting such affective responses as anxiety, doubt, and uncertainty. I may be conscious of these feelings and the needs from which they stem, or I may not be. Information needs at this level are pre-linguistic and volatile. I will not yet have words for them, and I am in no condition to actually ask a question that will result in an answer I can use to satisfy my need. At the next level up, my need becomes fully conscious. A *conscious* need is one that has acquired language, an explicit association with a problem, and the construction of a problem space. I know I have a need, I can talk about it, and I have a good idea it is a need for information, but my thinking at this level of need is unfocused and characterized by ambiguity. I will deliberately reflect on my need, but I may not yet be able to fully articulate it.

Under these conditions, I am likely to seek out and engage in conversation with various other people, including friends, advisors, and teachers in order to try out ideas and use their feedback to refine my thinking about my need. At its third level, my need for information becomes *formalized.* I can now make a formal statement of my need that expresses its aboutness. My need becomes objective in the sense that this statement is an object on which work can be done. I can articulate a question that is answerable. However, it is unlikely that I am thinking of how to ask my question in terms that a database, for example, can recognize and to which it can respond. This occurs when my need reaches the final or *compromised* level. A compromised information need is in effect a translation. My formalized need is now recast into system terms, so that it can be objectified. It is indexed and transformed into a query whose aboutness is mapped on the representative aboutness structure of an information retrieval system.

Much work in information science deals only with information needs at this level, yet there are good reasons for wanting to know more about how information needs are formed and articulated. It is important to understand, for example, that the four levels of information need do not represent a quantum-like reality, with each level of articulation clearly separate and separable from the others. Rather, they form a continuum from problem space to query space, with boundaries between levels more like those that mark day from night than those that establish legal borders. My problem space will be filled with visceral information needs, mixing in ways that both confuse and, on occasion, come together to provide a sudden flash of insight. At any given moment in my life some of these needs, for any number of reasons, will find expression while others go ignored. Some may rise all the way to a formalized level. Others may start in that direction and fall back again as other, more pressing needs become relevant.

Successful question negotiation, that is moving from a formalized to compromised need, requires direction and structure. Relevant communicative acts must be accomplished if final compromised needs are to accurately represent information seekers and what they want to do with the information they access. According to Taylor's research, librarians contribute to the creation of a query space by deploying five filters through which an information seeker's formalized need is passed. The word *filter* is an unfortunate term, since their respective names suggest a much more active process than distillation. Each filter represents a negotiating point that must be settled if the encounter is to come to a successful resolution. From the librarian's point of view, these filters represent steps in a conscious strategy used to ascertain and answer a user's question. From the user's point of view, they represent steps at which he or she can make and offer relevance judgments regarding the progress and outcome of the information-seeking process. As with Taylor's levels of need, these phenomena can be analytically separated, but they are not marked by discrete boundaries. Movement from one filter to another is not necessarily straightforward, and the librarian might very well be using two or more of at the same moment.

That said, let us take a look at the work of a reference librarian. The first filter involves talking with users to determine the content, structure, and limits of their questions. For instance, a librarian may use the results of preliminary searches to obtain relevance feedback from the user at this point in order to test hypotheses concerning the user's need. Dialog, whose goal is to negotiate and establish the user's criteria of topical relevance, is an essential part of this process.[6] The second and perhaps most critical filter determines the object and motivation of the user's search. Why does someone seek information? To what end will it be applied? If the first filter determines the aboutness of the user's question is about, this one is aimed at determining the aboutness of what is he doing and why. What is the nature of the contexts, both internal and external, that the user brings to the negotiations? Curiously enough, information seekers can often better articulate why they want information than they can describe what they want. In other words, they can intelligently discuss their problems even if they cannot articulate their needs.

These first two filters ensure the negotiation of relevance criteria, in other words, the criteria of judging the utility of retrieved information, and the manner and extent to which it is actually related to the user's problem. Later, as we shall see, the librarian will have the opportunity to exercise a prominent role as master of the system, but in these early stages, the user's view of the situation is, or should be dominant. It is entirely possible, if not likely, that the difference between the subject of a question, and its object and motivation will be a subtle one, and while the determination of each is logically discrete, these tasks will be accomplished more or less interactively and simultaneously.[7]

The goal of the third filter is to establish the personal background and characteristics of the information seeker. The user's status, experience, and specific, contextual relation to his or her question all come into play in order for the librarian and user to establish a mutual understanding of the user's situation and problem space. From where does the question itself arise, and what implications and shadings of meaning accompany it? How much does a user already know, and what kind and depth of information and information sources can he or she use? How urgent is the user's need?[8] Finding the answers to these questions often requires delicacy on the part of the librarian, and serious ethical issues turn on the problem of what should and may be asked. If the librarian is to be of any help at all, he or she must make certain judgments about the user.

Since the librarian is not in a position to demand information, the extent and kind of information a user will reveal must be negotiated. If users do not entirely trust the librarian or the system to understand and respond to their problem, they give away just enough to allow the system, including the librarian, to produce results that they will evaluate later in a private context. On the other hand, people now have the opportunity to seek out information from sophisticated retrieval systems without the benefit of professional mediation. A failure to be honest with oneself about what one knows and doesn't know is sure to result in a failure of the system to produce desired results. Like any process of social negotiation, the negotiation of questions in the context of a search for information can be marked by discontinuity, ambiguity, evasion, and cross-purposes.

The fourth filter belongs almost exclusively to the librarian, but usually requires consultation with the user. It enables the librarian to establish a relation between the user's question and the file from which an answer will be drawn, translating the query into a search strategy. The question, then, is interpreted and reformed in terms the information system can recognize and use. Thus, its success depends heavily on the librarian's knowledge of the constraints and possibilities of the system and on his or her ability to match the need to the system.[9] At this point the librarian becomes the user's representative. All of the issues and ambiguities regarding the representation of texts come into play in this act, and the librarian's efforts may fail for many of the same reasons that representation often fails. The librarian may be unable to represent the user's problem space to the system.

The fifth filter allows the librarian to negotiate the user's question through a process of answer specification. The purpose is for both parties to agree on what an acceptable answer will look like. The successful outcome of this negotiation may require users to alter their original ideas about what information may be available and useful and, in some cases redefine and reconceive their problem.[10] Since the above does not necessarily occur as a logical sequence of discrete steps, question negotiation is no different from developing and articulating a need for information. Both are characterized and signified by metaphors that stress the indeterminacy of the boundaries between the different conditions they identify. Different parts of the process of question negotiation occur at different times as librarian and user construct a definition of their mutual situation. In fact, it is entirely possible that the iterative and exploratory searches a librarian performs to establish the topic of the question and the nature of the topical relevance at play will also be the means by which answer specification is accomplished.

These five filters represent the means employed by librarians to initiate and guide the process of question negotiation. Information seekers must struggle more or less on their own to make the journey from visceral to formalized need, even with help and advice along the way. Their arrival at the gates of a formal information system, however, marks the end of one journey and the beginning of the next. It may be possible that this second journey can be conducted without the aid of a human guide due to advances in artificial intelligence and information technology and may someday allow us to dispense with the services of librarians, but these devices will still have to navigate the aforementioned filters. Either way, Taylor's work provides an essential foundation for understanding the source and nature of information needs and what happens when an information seeker asks a question of a librarian in order to satisfy those needs. But if this work represents what an information need and the act of seeking information looks like to a librarian, how do they appear to information seekers?

Information Needs and Information-Seeking

Carol Kuhlthau has investigated information seeking as a subjective process of constructing information needs. The central focus of her research is on the information seeker's experience. She wants to understand the psychological processes people use to construct and make sense of their needs for information.[11] If we are to understand what people are about when they experience a need for information and take action to satisfy that need, she says we must understand that the question they finally pose to an information system is merely the tip of an iceberg. People in need are about more than their queries and formal search strategies. If information needs arise from the roles people play in social life, then we must understand that a need for information represents some other, more complex, and more subtle condition than can be represented by a single question and its answer. By focusing on information seeker as a whole person, Kuhlthau explores information seeking as a contextualized process of constructing understanding from an initial state of uncertainty.[12]

From the perspective of the information seeker's experience, information seeking can be characterized by a series of six phases not unlike the steps undertaken by librarians to negotiate reference questions. The phases are not discrete steps in a logical sequence, nor are they separated by clearly marked boundaries; the completion of one phase does not in any automatic or inevitable way cause the next phase to be engaged. Kuhlthau describes the information seeking process as being indeterminate; it is entirely possible to move backwards as well as forwards. While each phase depends to an extent on the completion of the one prior to it, it is possible for someone to be partially engaged in any other phase at almost any time during a search for information. Finally, these phases do not represent a single discrete event that occurs within some limited or specified period of time. More likely, an information search will take place over an extended period of time, and events that occur during this time can and do affect the nature of a search and its purposes. The most important characteristic of the process to note, however, is that it incorporates cognitive, affective, and physical aspects within each phase. In other words, each phase is characterized by certain kinds of thoughts, feelings, and actions, which can make a difference to the information seeker's interpretation of retrieved information, to his or her interpretation of the information-seeking situation, and to the final success or failure of the search.[13]

Let us assume that in my life I encounter a problem that I cannot make sense of or resolve. It's very likely that for a number of days, perhaps even weeks, my information need will remain visceral. I am certainly aware of my problem, but it may take me some time to realize and understand that information will help me solve it. I may even believe that I already know, or at least ought to know, enough to solve my problem. However, at some point it will dawn on me that a large part of my problem is that I don't know enough to solve it. My information need has finally become conscious, and I can go to work on it. I can now begin to think about my problem. My anxiety may be lessened, but to say that I feel good would be to overstate the case. I know that I lack understanding, but my thinking is vague. I feel apprehensive, but I take action to *initiate* a search for information.

By completing this action, I begin the process of constructing a search for information. My task now is to actively identify and *select* a more or less specific topic and to begin to examine what seems to be information generally relevant to my need. My thinking centers on possible topics, personal constraints and criteria involving the nature of my problem and importance to me, and preliminary predictions about the likely usefulness of alternative lines of investigation. Note also that I am in the process of constructing a problem space. I feel optimistic about finally getting started. I am now ready to *explore*.

Exploration, however, can be quite troubling, and it is easy to pull back from the quest. I find that the more information I retrieve, the less certain I am about what I'm doing. On the one hand, I'm busy—locating apparently relevant documents; reading abstracts; comparing and contrasting competing facts, advice, and opinions. On the other hand, I'm not at all certain that I'm making correct use of those sources. Either the librarians are unable to understand what I need or I'm not expressing myself well, I don't know which. No matter what I do, I run into dead ends. My sense of doubt and anxiety is rising, and I can't seem to find an angle on my problem that will allow me a clear path to its solution. There are times that I feel like giving up the whole effort. Still, I strive to establish a consistent and coherent point of view. If I can do that, I'll know which information I have is relevant and which is not.

As I explore, if I am not careful, I will pursue too many possibilities, follow too many leads, judge too many different texts to be relevant, and end up with more information than I need. In other words, my sense of what I really need may be lost. If I am successful, however, I can now *formulate* a point of view that I can apply to my problem. This is a turning point. My uncertainty begins to fade and I feel confident that I'm on the right track. My thinking is focused and grounded in a solidly constructed and formalized need. I can make a clear and concise statement about what I'm doing and why, and my actions have a definite purpose to them. I can now make secure relevance judgments without fear of discarding something I need or keeping something I don't need. As I prepare to actually *collect* information, the transition is subtle and almost unnoticed.

While not finished with my search, I find myself already using information I've retrieved, and my interest in my project is growing. I feel as if I have a definite sense of direction and control over my sources of information, if not my problem, and that my actions are effective at retrieving relevant information. In Belkin's terms my anomaly is beginning to dissipate. At last I am ready to *present*. I feel relief as I bring my search to a close, confident that I have found information that I need. While the problem that motivated my need may not be solved, I think that my understanding of the informational aspect of the problem is complete. Already I know more than I knew before I began my search, and I am beginning to use it. Whatever compromises I have had to make with the system are the result of conscious decisions, and I am satisfied with the outcome.

Throughout the entire process my feelings have alternated among anxiety, optimism, confusion, and, if I was successful, satisfaction. My thinking was vague at first, but with progress it became more focused. My actions were initially hesitant, but became increasingly decisive. These feelings, thoughts, and actions that I experienced reveal a beginning and an end, but the process was not necessarily linear or logical. At some points, I felt satisfied, only to experience anxiety at others. My thoughts sometimes cleared only to be come cloudy again. My actions were sometimes increasingly rational strategic moves, other times almost random. Neither the information-seeking process, nor my feelings, thoughts, and actions regarding this process can be understood as inherently progressive.[14]

For example, if after considerable effort I really didn't think I'm any closer to a solution of my original problem, the feeling of disappointment and defeat could be quite sharp, leaving me convinced that there was no answer to my problem. Or perhaps during the time I was engaged in a search for information, my situation would change. This change could be as simple as a new interest that leads to a changed paper topic or as complex as a complete re-evaluation of my life that led me to rethink the nature and meaning of my relations with others. I may also have a number of information searches underway simultaneously, some of which are personal and others work related. At times I will consult formal information systems or rely solely on informal sources. Each search can interact with one another in ways that alters my progress in each. We are just beginning to realize the depth, variety, and complexity of the problems associated with the search for and use of information.

The notion that human personal reality is the unstable outcome of a concept of the information-seeking process. Its instability does not mean that it is beyond control, but there are elements with which we must work that must first be accepted as givens. Many of the problems we must solve arise from the constraints that life imposes. For example, the information sources to which we have access may be proscribed by legal, political, or cultural restrictions. But even if we are not be in a position to make of reality exactly what we choose, neither are we free of the obligation to make sense of what we find in the world. Interestingly, the tools we use to construct a personal reality, including information, action, cognition, and emotion, are also the materials from which such a reality is built.[15]

Information-seeking is just one of the means by which subjective reality is created and sustained. The aboutness of this process is embodied and manifest in the uncertainty arising from our need to make sense of the world and our relation to it. When facing uncertainty, we seek out information to re-establish certainty. Feelings of confusion and frustration are transformed into feelings of confidence. This is a matter of work done on reality in order to change it. Thus, the construction of certainty is a matter of constructing personal knowledge—a dynamic and holistic experience that engages one's feelings as well as thoughts in a context of action—and all combine in different ways at different times during a search for information to influence the nature and direction of the search and its outcome.[16] If the use of information changes an individual's state of knowledge, it is by means of a self-directed, active process that may be neither systematic nor orderly.

During the search for information and the construction of certainty, subjective subcontractors are also at work. The process of constructing certainty is accompanied by an understanding of what one is doing. This development allows an information seeker to refine and extend the topic of the search as an outcome of the information retrieved, such that he or she may begin to use information prior to the completion of a search, even as that use can alter the direction and purpose of the search. Information seeking is often a matter of improvising around a theme, which accounts for why different people searching for information on the same topic often arrive at different conclusions and use different information. Similarly, the need to explore in order to formulate an approach to a topic requires an encounter with unique information which is not congruent with what is already known, and judgments of its relevance are difficult to make. Information that conforms to expectations is reassuring, but its redundancy means that it contributes little to the satisfaction of a need. Information seeking involves a necessary tension between the familiar and the unexpected.[17]

Overconfidence may accompany redundancy. Anxiety may accompany the unique. Both can bring an information-seeking process to a premature end. A tolerance for the unique must be balanced with the use of some redundant information in order to arrive at a point at which the unique can be made familiar. This aspect of information seeking is particularly influenced by one's overall affective response to the process at a given point in time. To cope with the unique requires a response that allows expansive actions and a sense of play. Collecting and using information based on a secure formulation of a topic and its relevance criteria requires a response that leads to conclusive action. Conscious awareness and alteration of mood at different points during the search process is an essential tool of construction, but awareness and control are clearly affected by one's ability to tolerate the ambiguity that accompanies encounters with new information.[18]

A search process also involves a series of predictions based on an information seeker's experience, expectations, and understanding of his or her need at any given moment during a search. Because the search itself will change these conditions, an information seeker's predictions about what will be relevant and about what to do next are also likely to change during a search. Some predictions will not be confirmed, and others will provide the basis for further work. It is possible that a prediction once discarded may be revisited as new information leads to the re-evaluation of what has come before it. As a search progresses, predictions are based on more secure knowledge and greater confidence, and the approaching closure of the process is signified by an increasing number of confirmed predictions. As a whole, then, a successful search process, as a subjective construction, will be accompanied by an increasing acuity of interest on the part of the information seeker. When the goal of the process comes into sight and when the topic and its implications become clear, the information seeker gains confidence that his or her actions are having their desired effects. At this point, motivation and intellectual satisfaction increase as well.[19]

Sense-Making

These explorations into the nature of information need and use challenge the assumption that certain kinds of information activities, such as library service or information retrieval, have any value at all apart from their use. Even the suspicion that this challenge might have value raises some interesting questions. How useful is it really, to think about information as an object that inherently possesses value as a means of producing certain predictable effects? Does the notion that information is some *thing* that allows individuals to more effectively adapt to their environment impose unnecessary limits to our understanding of what information is and what it can do? Does information science assume, erroneously or otherwise, that given enough information and good enough retrieval, every information need can be successfully defined, and needed information always found and applied? Can the kinds of situations that require information really be treated so routinely?

These questions imply that at least some research in information science, and some information professional in practice, may be using a concept of information that is as normative and prescriptive as it is empirical and descriptive. Brenda Dervin, a scholar of communications challenges information science on the grounds that its project is based on two untenable and unexamined preconceptions. First, information exists independently of human thought and action, and its value lies in its power to describe and reduce uncertainty about reality. Second, information possesses an inherent order and organization that can be identified and manipulated.[20] From Dervin's perspective, these preconceptions limit information science to questions about what kind of information is needed to successfully respond to given situations, what is the nature of optimal information storage and retrieval, what are the most effective ways of responding to need, and how can we best educate information seekers regarding the use of information. These questions in turn lead us to ask whether the physical and cognitive metaphors that have guided our study so far can help us now. Perhaps not, unless we entertain an alternative view of the nature of information.

The conventional view of human nature that supports the preconceptions we identified earlier is that human beings create and use information to adapt and adjust to a constantly changing environment. In this view, the world is a place in which there is order, and, whether we know it or not, this order is complete and perfect unto itself. Information is the phenomenon that describes this order, and, to the extent that this order is not fully described, we say that our information is incomplete rather than suggest that it might be the world that suffers from this condition. This view overlooks the possibility that human beings are also creative creatures who make up the world they live in at least as often as they respond to it, and that the order we find in the world is, in fact, an order that we put there. We create the order we live in as our *self*-creating actions make old adaptations obsolete and change the conditions under which we act and interact with one another. From this point of view, the world and our information about it is inevitably and necessarily incomplete. The human condition is one of being incomplete, and this motivates our actions in the world.[21]

By failing to embrace this view of human nature, and to recognize that the world we live in, including information, is to some extent a human invention, information science may be unnecessarily limiting itself to asking questions for which it already has answers, or at least specified what those answers should look like. The metaphors that organize and discipline information science as an intellectual and social activity serve as powerful and enlightening guides. Both the physical and cognitive metaphors pose intriguing puzzles whose solutions have contributed to the solutions of theoretical and practical problems, but as we have seen, the demand for internal consistency that each demands allows some questions to be asked while excluding others. Both are evidently necessary in order to cope with information as an ambiguous theoretical object, yet the challenge raised by Dervin suggests that information possesses qualities that eludes both metaphors. There may be questions that we need to ask that lie outside the scope of what either metaphor paradigm will allow. Dervin argues that information science is on the edge of asking some very interesting questions, but that it fails because it fails to understand an essential aspect of human existence. It fails to consider the way in which an individual's life is constituted by his or her movement through space and time, and that the space/time continuum that makes up an individual's life is particular and unique.[22]

She compares the normative practice of information science, including the professional practice of librarianship, to that of medicine. In conventional practice, when an individual seeks treatment for an illness, the doctor's focus is on the disease. The rule is to diagnose, classify, and treat the disease with known and normatively prescribed methods. The norms that guide the treatment are based on documented knowledge about the given type of disease and the stage of its progress in a particular case. In effect, the patient becomes the object to which an instrumental technology of proven success is applied. The unique experience of a particular human being engaged in a confrontation with the unknown is reduced to a typical instance of a general situation, and in this process the patient, as a human being, can become relatively unimportant and sometimes goes unnoticed as well. Dervin's implied question is, why should we expect the outcome to be different for information services if they too treat an individual's information need as something apart from the individual?

Of course the patient needs to be cured of the disease, and the application of knowledge is necessary for a diagnosis and appropriate treatment; but people who experience diseases also need to make sense of what is happening to them. In some cases, this requires "information" that has little to do with knowledge and a great deal to do with faith, in others a recognition that knowledge itself is essentially unstable and temporary. Either way, we are obliged to make sense of the world and our relation to it. The central notion to this way of thinking about information and knowledge is that of *making*. Sense does not result from some mechanical process of introducing information into a situation that is not sensible. As the work we've examined on information seeking suggests, sense must be actively constructed, a process that engages an individual as whole. Thoughts, feelings, and actions are combine to make sense of the movement through space and time that constitutes the creation of a life. This movement sometimes requires that we adapt to what is given, and accept the order of the world, or its lack, as it is presented. On other occasions we must create and impose order on the world. Information has a role to play in both actions, but each action requires a different kind of information.[23]

Dervin argues that the ambiguities of "information" as a theoretical object should be considered in the context of sense-making, but this view implies an unconventional and alternative concept of information. She posits that there three different kinds of information. Information One describes reality. It is external to the self and can appear in many forms. These forms are represented by a continuum that includes direct stimuli from the world that manifest the innate structure of reality at one end, to formal texts of various kinds at the other end. In contrast, Information Two is constituted by the ideas a person has of and about the world. These are the structures and meanings an individual imputes to reality, in effect the combination of personal cognitive and affective notions one uses to describe the aboutness of the world to oneself. It is internal to the self, subjective and unique to a given individual.

We need to understand that Information Two is neither the only source of the anomalies that may generate information needs, nor is it an obstacle to be overcome in order to get to the Information One that will resolve anomalies. Information Two is both a legitimate source of creativity and a means of both adapting to and changing the world. In this context, the problem of information and its use is not a matter of determining what information I need from world to resolve anomalies I experience regarding my consciousness of and relations with that world. It is also not a matter of bringing these two kinds of information into closer alignment, but of negotiating the confrontations between them.[24]

In my unique space/time, reality is an outcome of the interaction between Information One and Two, between information from the world as I encounter it and information that I have already constituted by what I think and feel. The question is not where can I find information that will complete or fill the gaps in my Information Two. My information problem is one of making sense of reality when the confrontation between Information One and Two cannot be resolved. How then am I to construct a meaningful, sensible relation between the world and myself? The behaviors I undertake to construct this relation, the thoughts and feelings, and external information I employ to make sense of it constitute what Dervin calls Information Three. While Information Three can also take many forms, in each case it is an active phenomena. One approach, for example, is to screen reality for alternative input, and then use

Information Two as the basis of a decision about what makes sense to me. I might also, or alternatively, rely on my preferences regarding what I think reality ought to be. Or I might seek out the advice or opinion of someone else. In any case, my Information Three, my creative response to any situation I encounter as I move space and time, will be unique.

Because Information One cannot completely describe reality, which must be experienced rather than described, Dervin regards the purpose and utility of information as a relative concept. Similarly, each individual lives in his or her own reality (Information Two) and on that basis uses procedures (Information Three) to assess and make sense of the world (Information One). This means that the purpose and utility of information cannot be an invariant characteristic of information as a "thing," but is instead determined by its user. The same Information One serves different purposes for different users, and for different reasons, because each is moving through a unique space/time continuum.

Thus, information is not valuable in its own right, but is dependent on how it is used and by whom. To control myself, and my movement through space and time, I must make my own sense of reality and my relation to reality. Sense cannot be imposed, or given to me by someone else. I am the only one who can make sense of my life. I may not be able to make this sense just exactly as I choose. Certain aspects of Information One will be unyielding. I may wish them away, but that does make them go away. I may make sense of my situation in a way that I cannot sustain because of my unwillingness to accept what is. On the other hand, the sense I make of the world is conditioned, not determined, and I possess at least a relative autonomy to choose to make what sense I will.

This application of the cognitive metaphor has a number of implications for information science. It shifts our focus from the informative object to the interpreting subject. Instead of asking questions about what kinds of information are used to accomplish what kinds of ends, we have to think about what was communicated. What is learned? What sense is made? What is the outcome of using information? These questions in turn raise other questions about the *situations* in which information is created and used. What kind of situations are related to seeking and using information? How does the nature of the situation affect the kind of information used, and the kind of use made of it? What do different people do to make sense of different kinds of situations, and why? How do people actually become informed? How and why do they select or reject information to use? What role does information play in the creative response to the movement through space and time that we call life?

Metaphor and Method

The research of all three of the scholars we have examined in this chapter depends on a methodology that is grounded in empirical observation, but does not conform to conventional expectations and criteria of scientific method as usually applied within the social sciences. Taylor arrived at his conclusions by observing librarians at work in real-life situations, with actual users whose questions were entirely of their own conception. Laboratory conditions were not maintained, controls were not exercised, and the design can hardly be described as experimental. We cannot statistically generalize from it. It cannot provide the basis for the discovery of fundamental natural laws, and its validity and reliability can be called into question. On the other hand, for anyone who has ever worked as a librarian, its summary description and conceptualization of that work is both powerful and persuasive. It also represents a method by which the cognitive metaphor's posited World Two, the world of subjective knowledge, can be penetrated, observed and understood. Taylor's work might not be regarded as an exemplar in the Khunian sense, but it certainly provides a model for further research.

Kuhlthau's work also employed a qualitative methodology in order to observe, understand, and explain the experience of people seeking information. She too constructed and employed a means of studying people struggling with real rather than experimentally determined information problems. The phases of the information-seeking process she identifies were not hypothesized in advance of the study. Instead, they were inductively created, emerging from a bottom-up process of observation and interviews with the research participants. Both Taylor's filters and Kuhlthau's phases can be identified as discrete categories of reality or conditions that can be indicated or measured by means of a quantifiable characteristic. Both are careful, for example, about recognizing the indeterminate boundaries that separate one category or condition from another. Both implicitly grant that at a given moment for a given research participant, it may not be possible to determine, for example, the actual state of his or her need, or the information seeking phase the participant is in. Another way of putting this is to say that these are not phenomena that can be identified or described in a language of physical things.

Instead, the plausibility and truthfulness of Taylor's and Kuhlthau's assertions depend on two acts. The first act depends on the researcher conducting an honest and empirical investigation. While not hypothesis driven experimental work, both scholars set explicit rules for their observations, base their conclusions on these rules, and maintain an internal logical consistency. The second act, however, depends on the reader's recognition and endorsement of their observations and conclusions. In effect, it is as if Taylor and Kuhlthau are asking their readers, does this ring a bell? Do you recognize the explicated experience? The reader must act to verify the findings of the research.

This method exemplifies the need for metaphorical language in talking about certain human realities. The cognitive metaphor of "information" must embrace both the ambiguity of information as a theoretical object and the ambiguities of human experience. It ultimately posits "information" as an experience of consciousness. We may indeed participate in events that have very tangible and material aspects and consequences, but our experience of events inevitably remains an intangible and immaterial product of consciousness. These events, including the using of a text, will have tangible and material effects on our senses, but our interpretation of their meaning is inevitably a creation of consciousness engaged in a confrontation with material phenomena and conditions. The cognitive metaphor seeks to capture and explain an elusive reality. Its efforts and conclusions are compelling but the language of its discourse is troubling.

The notion of "structure," for example, is central to the cognitive metaphor. It appears in many places in the cognitive metaphor's discourse, as in "knowledge structure" and "image structure." Because cognitive models are themselves structures that represent these more fundamental structures, problems can arise. When we describe a knowledge structure as an anomalous state of knowledge, do we equate "structure" with "condition," and if so, is it a valid move? Or do we mean that "structures" can be characterized by their "state"? In addition, we have already identified a knowledge structure as phenomenon that can be attributed to both texts (things) and readers (people), not unlike communication, which connects their cognitive structures. A text is a limited expression of the thought of its creator that is linked to the thought of a user by means of communication. In effect, a text is a cognitive structure and a form of communication that links the minds of its creator and its user. Does this discourse begin to stretch "structure" as thin as the physical metaphor stretches "information" when in the context of the latter it seems as if every "thing" is potentially informative?

Other words and phrases raise the same problem. How, exactly, does one "construct" a problem "space"? Where is this space and what are the materials of construction? We noted earlier the actions of subjective "subcontractors" that illustrate the complexity of what was being described, but the reference is not without irony. Who are the laborers in this construction project? How much are they paid? When Taylor uses the word *filter* to talk about how librarians build models of a user's knowledge state and its anomalies, we called its possible reference to a distillation process "unfortunate." The same is true of many of the conceptual terms and phrases deployed by discourse within the context of the cognitive metaphor.

A phase, for example, might be called a step, albeit with slightly different implications regarding the nature of progress. A filter might be called a gate, again with slightly different implications. A knowledge structure is also called an image structure. The language of the cognitive metaphor is maddeningly imprecise, yet powerfully suggestive because it names phenomena that we experience. It speaks of intangible realities even as it cannot escape the desire to materialize its signifiers. As a result we read of the "construction of structures," the "building of models" and no matter where we turn we find consciousness confronting a World One, an Information One, or an information retrieval system characterized by material means of access to a collection of physical texts.

Conclusion

The physical metaphor drives efforts to study physical artifacts that represent information, while the cognitive metaphor directs attention to the study of people and their use of information as a cognitive phenomenon. Neither metaphor completely excludes a consideration of the other's emphasis, however, and their differences should not be allowed to obscure what the two approaches have in common. Each metaphor is concerned with the representation of information and knowledge embodied in texts of all kinds regardless of their medium of representation. Each insists that information science must be concerned with structures of information and knowledge in order to solve problems of information retrieval. Yes, there are differences. The physical metaphor suggests puzzles for information scientists to solve regarding the inherent and objective structures of texts and text databases and how best to match the contents of these databases with user requests for information. The cognitive metaphor, on the other hand, is concerned with the knowledge structures of individual texts and those of individual users and how these structures might be mapped on to one another.

The cognitive metaphor supports a science that is perhaps more of a *soft* behavioral science than a *hard* physical science, though both seek objective knowledge of the conditions and nature of information retrieval and use. The physical metaphor directs attention to the characteristics and behaviors of information as a thing, while the cognitive metaphor directs attention to the characteristics and behaviors of human cognition and the role of information in that process. In addition to these differences, the physical metaphor seeks the certainties provided by quantitative measurement of the attributes of information that can be described in the language of physical things, whereas the cognitive metaphor implies a need to explore empirical yet qualitative and nonexperimental research methods.[25]

The difficulty for "information" as a theoretical object is that it, must simultaneously manifest objective and subjective aspects. Ideas are more than their material expression, and we cannot assume that such expressions are literally the ideas they attempt to express. The connection between expression and idea is indeterminate and conditioned by facts, beliefs, values, and ostensible observations of reality. As we cannot assume a one-to-one correspondence, we cannot assume that the retrieval of a document necessarily, automatically, or inevitably informs. Even in the discourse of science, whose vocabulary is based on operational definition, the reported outcome of experiment is not the reality the discourse reports. The map is not the territory.

Reality, as known to a writer, must pass through a writer's cognitive structure—through his or her knowledge, beliefs, and values in order to arrive at the formalization of text. Text then, is merely a representation of the writer's thinking regarding its subject. It presumably expresses what the writer intended to say about a subject, but in all likelihood, it also implicitly, possibly unintentionally represents a writer's entire mind, including attitudes, feelings, and ideas about other subjects. It is both subtext and context, accessible only by reading between the lines and not usually formally indexed by any method.

But the process is not yet over. Once a writer's cognitive structure, or at least that part of the structure representing the thought from which a text is written, is manifest in the material and objective form of a text, it must again pass through a cognitive structure, this time one belonging to a reader. This passage is accompanied by an interpretation of the text according to what the reader already knows, believes, and values in order to be transformed into information. At this moment we confront again the categorical duality of information. Its objective nature is manifest in the formal expression necessary for its existence, which can appear as words, pictures, music, or any human-created artifact. Its subjective nature is manifest in the subjective apprehension and interpretation of this formal expression by a human mind. Thus, formal expressions of the human mind must be transformed into thought by a reader of those expressions in order to complete the process of creating information. And this reveals that information is finally a relation. Specifically, it is a communicative relation, mediated through text, between two minds. In a very real sense, information is the outcome of a collaborative creation of writer and reader.

On the other hand, until one actually writes a text, or at least speaks words heard by a listener, and thus physically places "information" into the world, there is nothing to talk about. Representation then, on which information retrieval is based, remains elusive, and the judgment of relevance, on which the uses of information are based, remains problematic. Nevertheless, what we encounter here bears a striking affinity with the problem posed by the nature of the sign and the relation within it between signifier and signified. The value of the sign, like the meaning of information, depends upon a psychological unity of thing and idea, of text and content, and upon the apprehension and interpretation of that unity, in the case of signs by a speaker of language, and in the case of information by a reader of texts.

Endnotes

1. Robert S. Taylor, "Question Negotiation and Information Seeking in Libraries," *College & Research Libraries* 29 (May 1968): 179.
2. T. D. Wilson, "On User Studies and Information Needs," *Journal of Documentation* 37 (March 1981): 9; and Robert S. Taylor, *Value-Added Processes in Information Systems* (Norwood, NJ.: Ablex Publishing Co., 1986): 34-38.
3. Peter Ingwersen, *Information Retrieval Interaction* (London: Taylor Graham, 1992): 131-133.
4. Taylor, "Question Negotiation," 181-182.
5. Ibid., 182.
6. Ibid., 184-185.
7. Ibid., 185.
8. Ibid., 185-186.
9. Ibid., 186-187.
10. Ibid., 187-188.
11. Carol C. Kuhlthau, "A Principle of Uncertainty for Information Seeking," *Journal of Documentation* 49 (December 1993): 244-245.
12. Ibid., 340, 344.
13. Ibid., 342.
14. Ibid., 342-344.
15. Ibid., 340-342, 345-347.

16. Ibid., 348.
17. Ibid., 348-349.
18. Ibid., 349-351.
19. Ibid., 351-352.
20. Brenda Dervin, "Useful Theory for Librarianship: Communication Not Information," *Drexel Library Quarterly* 13 (July 1977): 16-32.
21. Ibid., 21-22.
22. Reijo Savolainen, "The Sense-Making Theory: Reviewing the Interests of a User-Centered Approach to Information Seeking and Use," *Information Processing & Management* 29 (1993): 15-18.
23. Dervin, "Useful Theory," 25-26.
24. Ibid., 22-24.
25. Thomas J. Froehlich, "Relevance Reconsidered—Towards an Agenda for the 21st Century: Introduction to Special Topic Issue on Relevance Research," *Journal for the American Society of Information Science* 45 (April 1994): 126.

7

Representation of Information

Representation and Aboutness

One of the most difficult and persistent problems of information can be illustrated by a very simple question. What is this text about? This question may appear in a number of guises. We hear someone speak, then turn to a companion and ask, What is she talking about? We say something to someone and follow our statement with some variation of the question, Do you understand what I mean? So if language is not as straightforward as it first appears, then it stands to reason that what a text is about may not be immediately self-evident. Still, aboutness is at the heart of representing, organizing, and interpreting information, and we must resolve it if we are to retrieve and use information. In response, library and information science has advanced theories and built systems, yet an essential ambiguity surrounds the concept of aboutness and remains troubling for efforts to solve the problem of information.

The difficulties are revealed by turning the word on itself and asking, What is aboutness about? The self-referential quality of this question is a sign of the murky waters we are about to enter. Information science employs at least three different concepts of aboutness with regard to texts. Author aboutness implies that authors say exactly what they mean to say, and that they consciously use "content bearing units" to express the aboutness of their texts. This assumption provides the basis for natural language representation and automatic indexing.[1] There is a certain face validity about this approach for scientific language where words are defined by the operations that measure the phenomena they represent, but the case for making this assumption is less persuasive for other discourses.

Concepts in the social sciences, for example, are notable for their ambiguity, the humanities are even more problematic, and it is almost impossible to assign meaningful index terms to fiction and poetry. As the language of texts becomes more figurative, their aboutness becomes more difficult to determine and fix. Indexer aboutness implies an interpretation of a text so as to arrive at a summary or surrogate of its contents, and place it in a context of other texts determined to be about the same thing. This approach relies on the use of classification schemes and thesauri to organize information and to assign to texts descriptors that represent what they are about.[2] As the texts to be indexed display greater ambiguity of meaning, however, consistency of indexing, even with the use of standard controlled vocabularies, becomes more difficult to achieve.

Unlike author and indexer aboutness, user-related aboutness implies an ideal rather than a widespread practice and is employed to indicate the way the aboutness of text ought to be represented. This approach involves an anticipation of how a potential user would look for a text based on what he or she already knows, rather than try to summarize or represent its contents. Its goal is to begin with the familiar in order to enable the retrieval of texts that lead readers from what they know to what they don't know. An information need, after all, can only be constructed in terms of what a reader does not know.[3]

These different notions of aboutness make it a very confusing concept to discuss, and unfortunately there is more confusion yet to come because the divisions between the physical and cognitive metaphors of "information" as a theoretical object also come into play. For the former, the unchanging informative quality of a text constitutes its aboutness. It is the intrinsic subject of a text, independent of use or meaning, and communicated by content-bearing units whose meanings are relatively permanent and commonly understood. These qualities make possible the development and use of classification systems (such as the Library of Congress classification system and the Dewey Decimal system) and controlled vocabularies (such as the *Library of Congress Subject Headings* and the Sears list of subject headings). Thus, while meaning is closely related to aboutness, they are not the same. The physical metaphor insists that information retrieval systems can operate only on and with intrinsic aboutness by translating it into representations of a text's information-as-knowledge. In contrast, meaning concerns the extrinsic subject of a text that is related to the reason and purpose of its use.[4]

For the cognitive metaphor, aboutness deals more with more than the topical qualities of a text. Here, the relation between a text and the need for that text is central to determining what it is about. Given that a text may be topically related to a request and yet not useful for the problem that caused the request, or that a text may be useful in this way and yet not evidently topically related to a request, shows that here too aboutness may be different from meaning.

All of the above can be seen in a simple and conventional division in the notion of the word *about*. On the one hand, to say a book is about something is to make a statement regarding the topic or subject of the book. This idea clearly informs the physical metaphor's approach to aboutness, and it is expressed in all of the concepts of aboutness discussed above. On the other hand, there is a sense of the word *about* revealed by a question not infrequently posed in British films and novels. What's he about? Or, what's he on about then? These constructions are usually taken to mean What is he doing, and why is he doing that? A better statement of the cognitive metaphor's approach to aboutness cannot be found, and once again the unity of apparent opposites is revealed by the need to consider both meanings of *about* in order to have a complete understanding of the concept of aboutness.

As we have seen, the relationship between the two metaphors of information as a theoretical object is illustrated by various arguments regarding the reasons for using information retrieval systems. On one hand we can say that the purpose of information retrieval systems has little to do with answering questions, satisfying needs, or even resolving anomalous states of knowledge. Rather, its ultimate purpose is to retrieve texts that will help users of the system to do these things.[5] On the other hand, people do not go to a library to get books, they go to get what is in those books. The retrieval of information is not the final end of information seeking.

Regardless of the position one takes in this argument, to be of any help an information system must be able to retrieve texts whose content is about what a user needs. An information system must be able to make a connection between texts and the intentions of users, and to do so well, must confront and resolve the problem posed by the categorical duality presented by information as a theoretical object. For both texts and queries, it must discern *aboutness* in both of its senses. It must identify what texts and queries are about in terms that describe their contents in some logical, knowable, and objective manner. Without a topical match between text and query, the effort to connect a text and a user's intention for seeking a text ends very quickly; but even if this accomplished, the task is not necessarily completed. An information retrieval system should also be able to link what a text is doing and why, with what a user is doing and why. Once more, we can see both objective and subjective aspects of information. The words of the text and the query are physical entities that can be matched for the purpose of retrieval, but the meanings of the text and query are cognitive phenomena whose match will make retrieved information useful. Among other aspects, these situations demonstrate that the issue of whether information is a physical or cognitive phenomenon is not merely theoretical but intimately linked to the kinds of pragmatic problems both information scientists and librarians must solve.

Given that the quantity of potentially useful texts, even in a small library, far exceeds the capacity of human memory to manage, information retrieval systems are challenged to organize texts for retrieval according to what they are about and then accurately and adequately represent their aboutness. In practice this involves bibliographic description, classification, indexing, and abstracting. Improvements in practice, however, depend on finding answers to a set of theoretical questions regarding the best ways to determine and represent the aboutness of texts.[6] How *should* content be represented?

Such questions also confront ordinary information seekers employing natural language in a free-text search engine. They too, must make determinations of aboutness. Their queries not only represent need, they also provide the terms by which the search engine will organize the texts in the database for retrieval. While a match will presumably represent a text that is about the query, care must taken to ensure that a real connection between a user's intention and the content of the text does in fact take place. An objective match between search terms and index terms does not guarantee the satisfaction of a need. Two related but different judgments are needed if the goal of retrieving information for a purpose is to be achieved.

Another way of ascertaining this situation is to say that the use of any collection of texts requires the selection of a subset relevant to a user's need. This subset could be selected by examining every text in a collection, except that is prohibitively costly. Instead, retrieval must depend on a formal and logical representation of aboutness. By extracting some aspect of the original text or source of information, and having it stand in for the original, we can achieve that end, but the rules of extraction and their explanation must be are well understood. By creating and using a code whereby one object represents another object, there needs to be clear rules about how that code works. These rules, however, are unlikely to be free of ambiguity, nor are they especially likely to be easily grasped. One of the most common reasons for failure on the part of ordinary library users to take best advantage of the library's catalog is that they do not know how subject headings work or why they are assigned. Furthermore, any representation of the aboutness of a text is also necessarily a reduction, since a text's stand-in must always be less than the text itself. Even if the rules for selecting representatives are clear, the reductionist nature of representation means that some ambiguity will always be present.[7]

Representation engages a set of means by which one thing comes to stand for another, and the process of representing engages the attributes of disparate objects and concepts, socially constructed and sometimes idiosyncratic codes, and individual cognitive abilities.[8] The representation of aboutness inevitably results in a loss of information, and yet without this process and the loss it causes, systematic information retrieval is impossible. For example, a map is not the territory it represents but a guide to that territory; in this sense it signifies the territory.[9] Taken together, signifier and signified make up a sign that embodies an original referent, here "territory," which can be an object or concept. The map is the representation (signifier) and the territory is the original referent (signified). In order to make sense of the map, however, its users must first understand why the map was created. Second, they must know how to read the map's code in order to understand what attributes of the territory are represented. In this way, a map resembles an index term assigned to a document. Both are signifiers intended to represent a much richer and more complex signified. Both are like the tip of an iceberg, and for both the crucial question is how well do they represent what is beneath the surface.

Still, as useful as this metaphor is, there are differences between a map and an index term. A map generally represents a single territory in a one-to-one correspondence. In contrast, an index term is usually intended to represent the aboutness of a group of texts. Whereas its relation to what it signifies appears to be one to many, this appearance can be deceiving. Its assignment to a number of different texts is intended to imply that all of the texts to which it is assigned share a single common quality or topic that the index term describes. Putting aside for the moment problems of accuracy in assigning index terms to documents, we are still left with the question of whether it is reasonable to assume that all documents to which a term is assigned will be do the same things for the same reasons. Such representation may fail to embrace the full meaning of a text.

Of course, in operational systems it is common to assign more than a single index term to a text, but how many terms are needed to really cope with the meaning of a text, given that it depends to some extent on its reader? Aboutness for information retrieval becomes a matter of selecting and grouping texts in order to best match a need posed in a query by a user. While we may study the attributes, properties, and features of texts that will allow us to accomplish this goal, we still confront an essential indeterminacy. This condition arises not because the aboutness of a text cannot be finally determined, but that it can be determined in a variety of plausible ways. We may be able to create a really good map, one that will allow us to avoid the most well known hazards and find our way reasonably quickly to our destination, but it is still just a map, and it cannot guarantee that we will not encounter some surprises along our way.

Indexing is the means conventionally employed by information science to create maps of knowledge. It is based on the assumption that two interrelated operations are necessary in order to represent the aboutness of texts. The first involves an implicit or explicit conceptual representation of the contents of a text; the second involves the translation of this conception into the lexicon of a controlled vocabulary. Of concern to many is that the first operation is both fundamental and resistant to analysis.[10] The process of deciding what a text is about is one of the least discussed phenomena in information retrieval and arguably the least reducible to rule.[11] No matter how sophisticated and complex an indexing language or system may be, if the initial decisions regarding the aboutness of texts are wrong, then the representations derived from these decisions are likely to prove inadequate and retrieval of texts problematic. As we shall see, a variety of methods are available for making these decisions, but whether arrived at by human thought or the application of an automated algorithm, they all reflect the intuitive notion that aboutness arises from an interpretation of a text. The ambiguity of aboutness arises from the always more or less indeterminate relation between signifier and signified—between words and their meanings and between a text and its representation. The resolution of this ambiguity necessarily requires interpretation.[12]

The problems of interpretation for the purpose of information retrieval are twofold. First, indexers very rarely read entire texts in order to assign index terms, given the costs in both time and money. Automatic indexing systems, based on programed algorithms which instruct computers to identify and count words in texts, have been developed in an effort to avoid the ambiguities of interpretation that arise from human indexing. Either way, both approaches raise questions about the adequacy of interpretation and representation of aboutness. Second, they are attempts to predict yet another interpretation: that of the user. Users, however, are notoriously independent interpreters of information, and they may disagree with any a priori prediction of its aboutness after reading the retrieved text.

Nevertheless, pragmatic solutions that work reasonably well have been found. At a more fundamental level, however, lies the act of interpretation itself. To interpret a text we must not only summarize its contents. We must also arrive at some judgment regarding the meaning of what we read, and to offer a statement regarding what the text is about we must be able to explain and elucidate that meaning. The act of indexing, no matter how it is accomplished, implies a judgment regarding the meaning of a text. The assignment of an index term is conventionally taken as an indicator that the meaning of the text in question is in some way shared with the other texts grouped together by the same index term. On the other hand, the difficulties associated with isolating and examining the subject of a text suggest that we should be very careful about assuming that the subject of a text is either easily observable and inherent.[13]

What Is the Subject of a Text?

Determining the subject of a text is a question of interpreting its content and a matter of understanding its relationship with the reader. This condition introduces yet another a troubling note of relativism. Many, if not most, meanings assigned to a text by careful and reasonable readers will be predictable from its contents, so the problem of describing aboutness does not appear to be intractable. The range of meanings available for most texts, however, is still quite wide, and the need to interpret texts in order to establish what they are about raises problems for efforts at bibliographic control that ideally seek the certainty of unambiguous, one-to-one relationships between index terms as signifiers and the contents of the texts they signify. A full determining of the aboutness of a text requires that we distinguish functional representation of meaning from the mere description or summarizing of a text. Aboutness, as a matter of interpretation, is an extra-descriptive phenomenon, generated partly by the intrinsic subject of a text as signified by its content bearing units, but variable by user.[14] In other words, aboutness, because it is a construction of human interpretation, must be regarded as a cognitive phenomenon, but must also be based on the actual attributes of the texts whose aboutness is at stake.

This condition brings us to the second aspect of interpretation. When we say that analyzing the aboutness of a text is a matter of interpreting its subject, we are assuming some idea of what we mean by the word/signifier "subject." As might be expected, arriving at this meaning is not easy. What is the nature of the subject of a text? How do we conceive of it, and what problems do different conceptions cause for representation? Naively, we might view the subject as self-evident and contained within its various content-bearing units, for example its title. But to do so assumes that a word is an attribute of the thing it names, and as we saw with the map metaphor, this assumption disregards the ambiguity of signification—the relations between aboutness and meaning, signifier and signified, and the objective and subjective aspects of information.[15]

A different approach is to regard the subject as an idea, a concept, or an understanding. In this perspective, it is an intangible creation of thought about an object not an attribute of that object. Thus, a text may be about reality, but it is not the reality it is about. In fact, there may be as many "subjects" of a text as there are readers of that text. This approach thus privileges subjective over objective reality by essentially asserting the claim that the subject of a text is in the eye of the beholder—that it is what it is perceived to be. Unfortunately, its relativism tends to defeat not only representation but also all forms of shared meaning including ordinary human communication. In order to determine the "real" subject of a text, we would have to be able to access its author's mind, and if we could do that, there would be no need for the text itself.[16]

A third approach suggests that if different people apply the same means of analysis to a text and arrive at the same subject, then we can conclude that the subject is an objective attribute of the text. This idea rests on two pillars. The first pillar holds that ideas have universal, fixed properties, limited by essential categories of existence, which we may know or not, but which exist independently of their cognition. The second pillar is that the universal and fixed properties of reality, and the texts that represent ideas about reality, can be analyzed and discovered by means of a universal method, usually scientific in nature, whose conventions allow us to arrive at a special and privileged form of consensus about the nature of reality. This approach, however, obliges us to a view of reality which can lead, in effect, to a kind of absolutism that excludes alternative interpretations of texts. The extremism of relativism is avoided but at the cost of an equal error in the opposite direction.[17]

To avoid either extreme, a pragmatic approach to the concept of subject occurs in everyday practice. This approach recognizes that subject representation must above all be instrumental. Its purpose is to identify texts for retrieval, anticipate users' needs, and make it possible for them to find the texts they need. Central to this idea is that the concept of subject must serve as a means to the end of retrieving texts. The use of conventions based on consensus regarding the meaning of words and texts is employed to avoid relativism, but the notion of the subject as an inherent quality of a text is rejected. A subject is a subject *for* someone or some purpose, and if this is considered when representing the aboutness of texts, then room for multiple interpretations is introduced. The real difficulty here, however, is an inconsistency of subject term assignment that soon reaches a point where the meaning of the representation is lost as it leads to texts whose relation to one another is no longer clear.[18]

Each of these competing approaches reveals problems in the process of interpretation, and we might well ask if there is any reliable way to interpret a text that will arrive at a useful representation of its subject. An interesting attempt to answer this question is premised on a view of the world as composed of things which possess objective properties. These properties, in turn, exist independently of our perception of them so are not creations of our perceptions (not unlike the physical metaphor of information as a theoretical object). Texts, however, are things that present a special kind of theoretical paradox. On the one hand they express an author's subjective view of the objective world, and on the other hand they possess properties which are themselves objective and characteristic of the text as a thing. Given the points made in previous chapters, this paradox should sound familiar. It is an objective fact, for instance, that a text contains certain subjective statements that may either true or false. Nevertheless, these statements, taken as a whole in the form of a text, have the property of being potentially informative, and true statements can be made about them without the requirement that we judge the truthfulness of the text itself. The fact that we can make such statements about a text, based on its objective properties, means that we can represent its aboutness.[19]

Specifically, we can make three kinds of objective, true statements about any text. We can say something about the aspects of reality the text reflects. We can say something about how a text treats those aspects, and we can say something about a text's relation to other texts. And we can say that a text possesses certain objective properties that allow us to represent them. The determination of a subject, then, is a matter of evaluating and assigning priorities to the objective properties of the text based on an analysis of its informative potential, a compromise between relativist and absolutist extremes without the inconsistencies of a pragmatic approach. Still, problems exist. No pre-determined algorithm can reveal the essence of a text, and the properties of texts that are in play do not comprise a well defined or definable set. Their identification and meaning can differ given the purposes and points of view that are brought to them.[20]

We are close to describing an objective basis for the interpretation of the subject and the representation of the aboutness of texts, but the dilemma of the subjective remains. Even if a text has objective properties, those properties must first be engaged by a perceiving human being before they can be interpreted. In other words, at best we can say that while texts do have objective properties, they are perceived subjectively and that interpretation arises from a dialectic encounter between a text and its reader. If subject assignment is a predicate of properties which are themselves predicates of a text, then the determination of aboutness for a text must be regarded as a series of reductions based on some logical link between reductive steps where each step both signifies and loses more of the text's meaning.

If an indexer or indexing system and a potential user of information sufficiently share a common set of understandings about the world and the texts that address the world, so much the better. I once accompanied a friend on a trip to a bookstore that eventually led to us standing in front of a shelf of philosophy books. I saw a title that looked interesting, took it from the shelf, examined it for a few moments, and then turned to my friend. Handing him the book, I indexed it with the phrase, "You need this book." He took a moment to examine it and replied, "Damn, you're right." Of course what made this kind of indexing possible was a fair knowledge of the subject matter covered by the book, but even more important was an intimate knowledge of my friend's thought and work. In this case, our shared understandings were such that a means of describing the aboutness of a text that was absurdly reductionist, and clearly incomprehensible to anyone but the two of us, worked better than any formal, algorithmic, or summarizing method. As indexers and users move father apart, however, their ability and willingness to trust each other's subjective view of the same object becomes less certain. Successful representation of and access to information depends on shared interpretations of aboutness.

The Ambiguities of Interpretation

In practice, the problem of interpretation represented by the determination of aboutness is often resolved through pointing. This solution directs a user to material, and its goal is to put the user in a position to judge the relevancy of texts rather than guarantee as an outcome the retrieval of relevant texts. This end can be accomplished by a variety of methods, all of them involving four basic steps: identifying an author's purpose, weighing the relative dominance and subordination of subject concepts, grouping or counting concepts, and applying standardized rules of selection.[21] By these methods, subsets of text sets are created so that needed texts and users can be brought together in a reasonably efficient manner. Information retrieval systems can and are sometimes based on extremely complex systems of language and meaning transformations, but in the end they all accomplish a rather fundamental and deceptively simple task. User needs for information are converted into queries, the concepts and ideas of texts are converted into representations, and the representations are matched against the queries. When the former match the latter, a text is retrieved.[22]

The trick to accomplishing this task lies in appreciating the context within which it occurs, and understanding which attributes of this context are relatively unchanging and which are influenced by culture, history, and other sources of interpretation. For example, the objects to be retrieved, as we have noted, are relatively fixed in nature. Texts are not going to rewrite themselves. The meaning of texts, and their representations, however, are not so stable, nor are users and their purposes for seeking out and retrieving texts. In order to fashion information retrieval tools that can cope with these changeable and changing attributes of information retrieval context, we have to study the nature of texts and their use over time, understand the changing relations between authors and users, and identify methods and tools of retrieval suited to individual purposes.[23] Given the need to accurately and adequately represent texts for the purpose of their organization and retrieval, we need answers to the following questions. How many elements must be extracted from a text in order to represent it? Which ones? Should they be extracted in their natural form or translated into a controlled and standardized form? Should we generalize individual concepts?

Successful information retrieval depends on a user and the texts sharing a common system and code of representation. Given this condition, the optimal answers to the above questions remain elusive.

The best answer to the first question will depend on the number of elements needed; to the second, whichever are useful to the user; and to the rest, whatever forms are consistent with the user's abilities and requirements.[24] Each answer suggests that the search for a perfect means of representing and retrieving texts is akin to the search for the Holy Grail, worthy of pursuit, but unlikely to be found. Even if these contextual puzzles of representation and retrieval could be solved, one problem remains: the indeterminacy of language itself.

Consider the following three signifiers: "velocity," "table," and "fast." Each signifier is related to a signified and an interpreting code, and the relations stand as signs. Each sign poses more or less difficult problems of interpretation. "Velocity" signifies a relation between distance and time which can be expressed by the mathematical formula $v=d/t$, where d=distance and t=time. "Velocity" can also be represented as a vector in a two dimensional space (Figure 7.1).

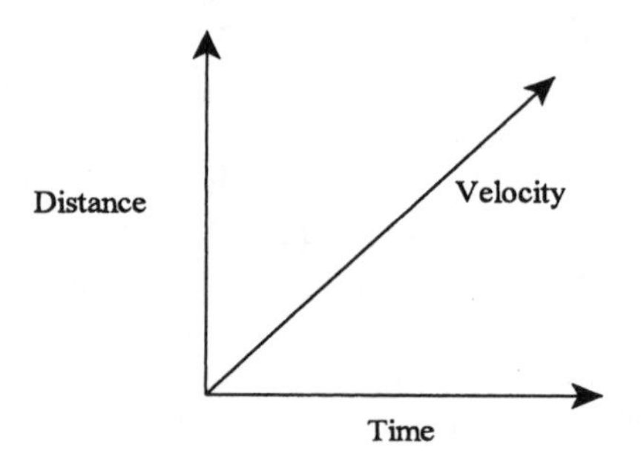

Figure 7.1 Velocity Vector

The code by which the signifier is related to what it signifies is operational, that is, velocity is defined a priori as a mathematical relation between distance and time, two concepts that are themselves defined by the operations by which they are measured. To follow what we discussed earlier, there is a great deal of certainty in this sign and not much information, as there is very little choice in the range of possible messages the signifier "velocity" can convey. Scientific discourse is typically composed of this kind of language, and its texts are relatively easy to interpret. They may be difficult to understand if one lacks a knowledge of the science they address, but from the perspective of information retrieval, the operational codes on which their language depends make them relatively easy to interpret in terms of aboutness. The kind of language they employ is deliberately constructed and employed to reduce ambiguity of meaning and make objective statements about the world.

Now consider the signifier "table." Clearly we're on different ground here. What does "table" signify? To being with, we must deal with the ambiguity "table" creates as a homonym. I might use this word to refer to a mathematical or scientific chart or a piece of furniture with a reasonably large flat surface supported by a leg at each corner. Even if I eliminate this ambiguity by specifying that I mean to signify the latter kind of table, there is still the question of what kind of "table." It could be Shaker style, mission style, French provincial, Victorian, or Oriental. Its size might vary and it might be made of oak, maple, or teak. In order for the signifier "table" to clear the hurdles of ambiguity associated with its possible meanings, we need to employ a code that identifies tables by the class of table to which they belong. Texts that employ words of this kind must also supply the qualifying and descriptive terms necessary to allow us to determine what kind of table the text is about, and we must match this supporting infrastructure with the vocabularies of our aboutness assignments or risk, on the one hand, failure to retrieve desired texts and, on the other, the retrieval of irrelevant texts. This is a difficult although not impossible task, and we have successfully developed and employed thesauri of controlled vocabularies for a wide variety of scholarly and everyday discourse that uses signifiers like "table" to convey information about the world.

Finally, let's take a look at "fast." It seems as if almost immediately we find ourselves in trouble. What can "fast" signify? To say that something is fast connotes that it is quick, speedy, swift, rapid, or, at least, not slow, but the trouble is all of these words define each other and, unlike velocity which can be measured with objective certainty, they imply a subjective judgment. To say that something is fast is to say that it is quick, but to say that it is quick is to say that it is fast and, in any case, in someone else's judgment, it might be slow. Before we can come to any common agreement on whether or not a thing is fast or slow, we must first establish a common frame of reference by asking, Fast relative to what? A day can appear to move with the slowness of cold molasses, yet we know that the earth is rotating at several thousand miles per hour. We can describe someone as "fast" and limit our meaning to speed with which a particular act is consummated, but that hardly does justice to the range and subtleties of meaning that are possible within the word.

The problem with words like "fast" is that the code needed to interpret their meaning is neither readily apparent nor determined. It is not difficult to imagine situations in which the codes needed for effective communication must be negotiated on a case by case, perhaps word by word, basis. Diplomatic treaty and labor contract negotiations are prime examples. Often the outcome of a legal case will depend on a negotiation regarding the code by which the relevance of facts will be judged and the outcome of the case resolved and, while legal principle and precedent are helpful, they are not determining. In fact, courts invite plaintiffs and defendants to advocate not only for their positions, but also for the codes by which their positions and the evidence will be interpreted.

Fortunately, many other texts, including those of interest to information scientists and librarians, are based on the intention to communicate a single message and so employ codes that allow us to classify them by the type of message they convey. To create systems of classification and control, however, we must first examine the nature of particular texts and their use. We must understand the relationships that exist between authors and users of texts, have a working concept of the subject of a text, and bring to bear the components necessary to construct retrieval tools suited to individual purposes.[25] Finally we must be prepared to cope with the way users of information and their purposes change over time and how that affects their interpretation of meaning.

Perhaps the best way to summarize the problem of aboutness is to refer to René Magritte's famous painting of a pipe on which he had written, "This is not a pipe." We could test the truth of this statement by attempting to smoke it, but the painting would go up in smoke, resulting in the permanent loss of a work of art, and likely an arrest for arson. Just so, every classification number or index term assigned to a text to represent its aboutness also declares, "This is not the text." Instead, it declares itself as a device, chosen according to a code that will allows us to retrieve texts on a given subject as long as we understand the relation between the device and the subject it represents. However, though we would very much like our retrieval tools to produce consistent, reliable, and valid results, value judgments prove difficult to avoid and at times more difficult to make.

For some signifiers like "velocity," the task is fairly straightforward. Given a group of authors and users for whom there is widespread agreement concerning its meaning, there will be few difficulties judging a text to be about velocity and relevant to a user's query about velocity. Even "table" is manageable if we are alert to the its different possible meanings and to new styles and fashions of furniture. Words like *fast*, however, pose serious obstacles for information retrieval that are not easy to resolve, since we have no certain way of knowing in advance which code to use while interpreting either texts or queries. It is for these reasons that scientific texts are relatively easy to control and retrieve, texts in the social sciences pose more problems, and texts in the humanities often remain elusive and difficult to adequately and accurately classify. The best access term for poetry remains the poet's name.

Each of these words—*velocity*, *table*, and *fast*—exemplifies the difficulty of determining an objective standard of utility for all information retrieval situations and the need to exercise a value judgment about relevance. The usefulness of a particular representation depends on the extent to which the representation is relevant to its subject, which in turn means determining which features are relevant to that judgment. In the end, an adequate and accurate representation should make it possible for users to retrieve only those texts that are relevant to their query. The retrieved texts should be about the user's need for information. The dilemma is that aboutness is just as elusive as relevance. For example, how should I describe what Magritte's painting is about, and which features of the painting are relevant to this judgment? Although clearly a representation of a pipe, the painting does not seem to be about pipes. In fact, the object represented by the painting seems irrelevant to a judgment of what the painting is about. Magritte could have chosen any material object as the subject of his painting; the key to its meaning lies in his commentary.

Note that I've made a value judgment about the painting based on my interpretation of a number of complex cultural codes regarding modern art, and that I've decided its most relevant feature is about the relation between the visual image it portrays and the written comment it includes about that image. So far, so good, and possibly objectively justifiable, but let's consider two possible subject terms that might be assigned to Magritte's painting: "representation—ambiguity" and "representation—certainty." I can, with some justification, assert a claim that his painting is about the inevitable ambiguity of representation, that it portrays the dilemma posed by the relation between object and representation. Or, I might assert that he portrays the certainty that object and representation are not and can never be the same. Which value, ambiguity or certainty, is most relevant to the "text" that is the painting? Even more important, which of these two subject terms would best serve someone seeking paintings that deal with problems of representation in modern art? Suppose this person was confronted with a record of

Magritte's painting that used both terms to describe it. How would the purpose of the search affect her judgment of the relevance of the painting to her need? Which value, ambiguity or certainty, will prove more relevant to her search, given that we cannot know this advance? Such value judgments like these are inherent in most information problems. We want to know, and need to know, what information is of greatest value, and in knowing, we want to retrieve that information.

Our task now is to examine a number of different systems of rules and how they govern the determination of the aboutness of texts. It is important to remember that these rules operate in exactly the same way that rules do in any game. Where the game is the representation of texts for retrieval, the rules should be able to resolve disputes regarding how the game is played. The existence of rules, however, means that somewhere there exists a rule maker who can change the rules and, thus, the nature of the game.

Indexing

From one perspective, the acts of retrieving and indexing texts can be described in the same way. Each is a matter of selecting and grouping. At its most general level, indexing is concerned with how to *best* select and group texts from a collection of texts to *best* match a need posed by a user's query. The physical metaphor suggests that the best way of doing so is to examine particular objective characteristics. What attributes, properties, features, or data elements of texts are relevant to the interests and retrieval purposes of users? Which of these attributes will best allow us to select and group these texts so that they may be more easily retrieved in response to a query? Indexing theory and research seeks to study these questions and predict their answers.

Indexing is a practice based on the logical and semantic attributes of texts, and the words that comprise them. These attributes in turn make possible the identification and construction of groups of texts. An index term or classification number indicates or stands in for the content of a single text by naming a group of texts with which the term is associated. The essential purpose of indexing, then, is to signify identifiable semantic features of texts in order to group them together on a logical basis and, then, name the group according to the characteristic attributes of its members. A theoretical understanding and explanation of the problem of indexing must consider two central variables. First, the nature and media of the texts to be indexed, as well as their indexable features, must be examined. Different kinds of texts will manifest different indexable features. This condition has implications for decisions regarding appropriate processes and methods of indexing, and for the structure, organization, and display of index records. Second, the nature and purposes of the collection of texts to be indexed and the audience served by a collection of texts must be considered. The characteristics and behavior of these variables are of interest both to scholars seeking a general understanding of the indexing problem and to practitioners who must actually index collections of texts for particular users.[26]

This accounting of variables, however, reveals another inescapable problem. To gain an understanding of the nature and purpose of collections, texts, and their audience requires the penetration of cognitive secrets. Behavioral science suggests some methods by which we might do so, but how can we tell what the author meant when he wrote, the indexer when she interpreted and assigned a descriptor, the selector when he chose, and the user when she read? Intention and its essential indeterminacy is involved in all of these actions. As a science, indexing's achievements are limited, even though prescriptions are abundant. There are some fairly confident conclusions about what works and what doesn't but, as of yet, little real understanding of why things work or don't. Much of the knowledge we use to guide indexing practice derives from history and practice.[27]

The goal of indexing is to help readers find the texts they need when there are too many texts for them to examine them individually. Index terms should be useful indicators of the subjects of texts to which they are applied. The problem is that usefulness is not an objective quality. The usefulness of a text, for example, can only be judged with respect to the person evaluating that text. In addition, the time, place, and manner of evaluation can make a difference to judgments regarding a text's usefulness. The topic of a text, like the particular objective features on which it is based, can be identified and it will not change. The meaning and value of the text, however, can only be determined by subjective and changeable criteria. Given that the act of indexing is really an act of prediction based on the indexer's judgment regarding the meaning and value of a text to imagined possible readers, it is likely to remain a dependable if inexact science.[28]

A second not so simple goal of indexing is to impose order on information chaos. To achieve this goal indexing must create an objective means of compensating for the essential subjectivity that characterizes judgments regarding aboutness. Human judgment must be transfigured into a system of logical rules and applications that allow at least the pretense of objective outcomes regarding the assignment of index descriptors, that is to say words, that identify the aboutness of texts and their meanings. This is a necessary condition for at least two reasons. First, the development and application of any logical system of indexing will contribute to the control of a constantly expanding body of information and texts, even if it fails to accurately and adequately represent every document. Second, we must justifiably maintain the verisimilitude of the fiction that we have objective control over our own creations if we are to have any confidence at all in our systems of information retrieval. To a great extent, the effectiveness of an information retrieval system is grounded in the belief that it works in ways we expect it to work.

These expectations are present in many of the affirmative definitions of indexing that appear in information science literature. The notion that indexing is a matter of analyzing textual content and expressing that content in the language of an indexing system is comforting but not entirely convincing.[29] The problem, of course, lies in the deceptively simple phrase "analyzing textual content." There is an implied expectation and confidence that this process can be carried out in a straightforward manner that overcomes, if not eludes, the subjectivity of interpretation. For the time being, however, we will put this matter aside and consider the notion of indexing on its own terms.

The central achievement of indexing practice and research is the development of controlled indexing languages, also known as controlled vocabularies. Controlled vocabularies are composed of predefined terms, that is words or phrases, that are sanctioned for use by indexers as they choose terms to represent texts. Natural language, or the vocabulary we use everyday to express our feelings, our thoughts, and our intentions, is incredibly rich and nuanced and capable of an extremely wide range of uses, but it can also be quite ambiguous. Of course, this ambiguity is precisely what makes similes and metaphors so interesting and contributes to our enjoyment of language, but this condition causes all kinds of difficulties for information retrieval. Conversely, a controlled vocabulary is just what its name implies. The depth, richness, and ambiguity of ordinary language, as well as the subtlety of meaning such language can convey, is sacrificed in favor of a limited number of words deliberately chosen for their ability to convey the essential aboutness of texts related to a particular subject. Alternatives are reduced in order to establish and confirm an expectation that the use of a given word in a formal query of a database or catalog will reliably retrieve the texts that one expects.

Controlled vocabularies possess many other advantages with respect to information retrieval. The content of texts added to a collection can be explicitly linked to the content of texts already in the collection.

Controlled vocabularies also promote and enhance the likelihood of the consistent representation of subjects by indexers and searchers, and their appropriate application avoids the dispersion of related material. In short, they are a great help in creating aboutness-based groups of texts. We now have at our disposal a number of different controlled vocabularies. Some are general while others are subject specific. But which of these vocabularies provides for the most effective retrieval and under what circumstances? Through which empirical methods can we best explain, control, and predict the behavior of indexes and indexing languages?

To answer these questions, information scientists have designed experiments that quantitatively measure the effectiveness of indexing. These experiments typically focus on the extent to which the use of different indexing languages affects the performance of information retrieval, the extent to which changes and improvement in indexing languages affect the improvement of performance, and the extent to which the cost of the improvement is worth the benefit derived.[30] In other words, it is assumed that the physical metaphor applies to the study of indexing languages, allowing a kind of research that isolates individual characteristics of indexing languages as variables that can be manipulated, singly or in controlled combination, so as to determine which indexing system works best under what circumstances.

There are a number of characteristics of indexing languages that can be studied and compared with respect to their impact on retrieval performance.[31] The structural and syntactical features of indexing languages can be studied and compared. Structurally speaking, the nature of an indexing language's hierarchical arrangement and especially its faceting which further defines the aspects of the represented concepts can make a great difference in the language's retrieval effectiveness. Different vocabularies are controlled in different ways, and these conditions are reflected by the specific words and phrases used as concept descriptors, the relations the vocabulary establishes between its descriptors, and the different means by which the semantic facets its descriptors are identified and organized. An indexing language may also be structured to assign a rank value to a term applied to a text as a matter of indicating the importance of that term with respect to the aboutness of a text. The use of this feature, called term weighting, can have a measurable influence on information retrieval.

Syntactical characteristics that can be explored include concept links and role indicators. Concept links express relationships between words in texts that represent and reveal meaning in ways that must be accounted for in order for indexing to be adequate. Role indicators, commonly sub-headings, specify the role of a word as a modifier of a concept in context and must be accounted for in order for indexing to be accurate. A Boolean search that links *venetians* and *blinds*, for example, might yield texts on either blind Venetians or venetian blinds when only items on one of these subjects is usually desired.[32] So which term modifies which? Different indexing languages possess varying syntactic features that affect their retrieval performance.

The next two sets of general indexing language characteristics involve the exhaustivity and specificity of a language. Exhaustivity of indexing refers to the extent to which concepts expressed in a text are actually represented by the descriptors assigned to a text, and it has two forms. Viewpoint exhaustivity is a matter of the extent to which the concepts of a text and their particular aspects can be represented by the indexing language. Importance exhaustivity is a matter of how important a concept must be before it warrants the assignment of an indexing term. Rules and policies regarding importance exhaustivity may vary by subject area and the purpose of the indexing.[33] The second characteristic, specificity, can also be studied with certain difficulties kept in mind. While one vocabulary may be consistently more specific than another, more commonly one vocabulary is more specific regarding some subjects while another vocabulary is more specific regarding others, depending on the purpose for which the vocabulary was designed. Overall comparison can be elusive, but it is possible to compare indexing languages on the basis of the generic level at which concepts assigned to indexed texts are expressed.[34]

Two final and very closely related characteristics of indexing languages that are often studied are manifestations of the behavior and decisions of indexers. Regardless of whether indexing is conducted by human beings or machines, issues of correctness and consistency continually arise. Indexing correctness is clearly central to the quality of information retrieval, and two broad types of mistakes are possible: omitting a term that should have been assigned to a text, and assigning a term that should not have been assigned. The first mistake fails to represent aboutness, the second misrepresents it. Two sets of questions can be asked regarding correctness. First, is the indexing complete? Are all the available appropriate terms assigned to texts? Of all the texts in a collection that should be assigned a term, how many are actually assigned? The second set of questions involves the purity of indexing—a mirror image of completeness. Of all the terms that should *not* be assigned to a text, what proportion are correctly rejected? For all the texts in a collection that should not be described with a particular term, what proportion are correctly treated?

Of course, underneath each of these questions lies the fact that a judgment regarding correct and incorrect can often be resolved only by a consensus among knowledgeable indexers and users, while consensus, though usually reliable, can be wrong. Of correctness and consistency, then, former is the more important. If indexing is correct, consistency generally follows. Indexing may be consistent, however, and not be correct. Still, there are ways to measure indexing consistency. We can compare the terms assigned to a text by two or more indexers and we can compare the texts to which terms are assigned by two or more indexers. Unfortunately, research suggests that indexing is not especially consistent.[35]

All of the characteristics of indexing languages discussed here manifest themselves as objects that can be physically manipulated to change the outcome of information retrieval performance. The quantitative extent of the manipulation and its effect on retrieval performance can be measured; and so, at least to that extent, the nature and behavior of indexing languages can be regarded as physical phenomena whose reality can be understood within the context of the physical metaphor of information as a theoretical object. There are some problems associated with this research, however, that point again to the categorical duality of information. Idiosyncratic elements can be and often are introduced in real world information-seeking situations through the specific nature of individual queries. The performance of an information retrieval system may be conditioned more by the idiosyncracies of a query and the person who poses it than by the objective characteristics of the indexing language employed by the system. In the laboratory situation sustained by this physical metaphor, queries and their meanings can be controlled, but there is still some room to doubt whether the results of such experiments will still be meaningful outside of this controlled environment.

Indexing and indexing languages are phenomena that manifest a number of thing-like features that are derived from the thing-like features of texts and languages and words. Still, there are limits to what we may be able to discover about indexing by the use of this physical approach. Indexing also possesses a cognitive element related to human decisions concerning the aboutness and meaning of texts and words, and the relation between indexing language characteristics and information retrieval performance is not clear cut. The wide variety and range of possible interactions between indexing characteristics are complex. This condition makes it difficult to study indexing by means of the classical experiment in which all possible variables are accounted for and controlled. An additional problem is that many of the important variables may not be quantitatively measurable. Solutions to some of the most important research puzzles and design problems may have to be sought by means of qualitative research.[36]

The assignment of cause and effect regarding the relationships of indexing to retrieval performance is difficult to do and, conceptually, a source of frustration in the development of indexing theory. It is not especially difficult to identify the variables that affect retrieval performance, but given the ambiguities of language and indexing, the identity of the variables and their relations that make the most difference remains elusive. Ultimately, information retrieval performance is a matter of agreement between an indexer, whether human or machine, and a user regarding the meaning of texts and words. Once again we are in the realm of signs, their value, and their interpretation. Design features of systems can be studied and manipulated to enhance the possibility of agreement, but we must face the fact that in the end human judgment regarding the code by which signs are interpreted may be more important for maximizing this agreement than the physical manipulation of indexing and retrieval system features.

Conclusion

By now you might be tempted to conclude that any attempt to represent the aboutness of texts is intractable at best. You might even go so far as to conclude that information retrieval is no more certain than gambling. True, the ambiguity of aboutness makes the systematic classification of texts a process of doubtful outcome and meaning; and if our goal is to achieve a perfect system for ordering and retrieving texts, your conclusions might be justifiable. Representation, when measured against reality, will always be judged incomplete, but any attempt to access reality without passing through representation is for all intents and purposes impossible. Fortunately, perfection is neither required nor desirable for the purposes of information retrieval. Within the limits imposed by the ambiguity of aboutness, less than perfect but effective solutions to the problems of retrieval are possible. Optimizing the representation of aboutness is a goal that information science has set for itself, to which end a number of competing approaches are available, each with its own strengths and weaknesses.

The origins of this effort can be traced to traditional concerns of bibliographic control, especially the practices of indexing and classification. The essential purpose of bibliographic control is to make access to information possible by bridging the categorical duality gap. It accomplishes this goal by logically associating representation with text and objects with their meaning. Bibliographic control is about making texts easier to find by grouping them together by subject. The problem is how do we get from the text and its ambiguity of subject, aboutness, and meaning to representation characterized by objectivity, accuracy, and adequacy and then, by the use of representation, to retrieval at which point the intentionality and purposes of authors and readers will meet and do as they will? The first step in this direction is to recognize that the purpose of bibliographic control is limited to representation for retrieval rather than for communication.

Avoiding the implications of information retrieval as a communicative situation is not entirely possible, but without reducing the complex problem of representation to a simpler form, we are again faced with the difficulties of achieving perfection. The need to organize and access information existed as a human problem well before it became a puzzle for information science, and solutions to this problem have been driven by a pragmatic need to do something, even in the face of obvious costs. A common element of these solutions is the manifestation and prescription of rules of representing aboutness by means of indexing and classification. Sometimes these rules have been based on entirely a priori reasoning and convention. Sometimes they have been based on empirical observation and theoretical assertions. In any case, these rules have been systematic and grounded in logic and the logical relations perceived to exist between concepts communicated by texts.

The next chapter will examine two experimental information representation and retrieval systems. One is grounded on the physical metaphor and the other on the cognitive. Neither system is in widespread use for a number of reasons not the least of which is the vested interest and prior investment of commercial firms in established methods and technologies. Nevertheless, both are of interest because they represent ideal types. Both nicely embody and illustrate the two metaphors of concern to us here as each serves to signify, through their design and operation, each metaphor. As practical systems, each reveals both the strengths and weaknesses of each metaphor's approach to signifying "information" as a theoretical object.

Endnotes

1. Peter Ingwersen, *Information Retrieval Interaction* (London: Taylor Graham, 1992), 50.

2. Ibid., 51.
3. Ibid., 52; and W. J. Hutchins, "The Concept of 'Aboutness' in Subject Indexing," *Aslib Proceedings* 30 (May 1978): 176-179.
4. Clare Beghtol, "Bibliographic Classification Theory and Text Linguistics: Aboutness Analysis, Intertextuality and the Cognitive Act of Classifying Documents," *Journal of Documentation* 42 (June 1986): 84-85.
5. Stephen E. Robertson, "Between Aboutness and Meaning," in *The Analysis of Meaning: Informatics 5*, ed. M. MacCafferty and K. Gray (London: Aslib, 1979), 204.
6. Hutchins, "The Concept of Aboutness," 173-176.
7. Ingwersen, *Information Retrieval*, 51.
8. Brian C. O'Connor, *Explorations in Indexing and Abstracting: Pointing, Virtue, and Power* (Englewood, Colo.: Libraries Unlimited, 1996), 1-2, 19.
9. Ibid., 26.
10. F. W. Lancaster, *Vocabulary Control for Information Retrieval*, 2nd ed. (Arlington, Va.: Information Resources Press, 1986), 3; and Bernd Frohmann, "Rules of Indexing: A Critique of Mentalism in Information Retrieval Theory," *Journal of Documentation* 46 (June 1990): 82.
11. A. C. Foskett, *The Subject Approach to Information,* 4th ed. (London: Bingley, 1982).
12. M. E. Maron, "On Indexing, Retrieval and the Meaning of About," *Journal of the American Society for Information Science* 28 (January 1977): 38-43.
13. O'Connor, *Explorations*, ix-x, 47-51.
14. Ibid., 147.
15. Birger Hjorland, "The Concept of 'Subject' in Information Science," *Journal of Documentation* 48 (June 1992): 172-173.
16. Ibid., 173-176.
17. Ibid., 177-179.
18. Ibid., 179-181.
19. Ibid., 182-183.
20. Ibid., 183, 187.
21. S. D. Neill, "The Dilemma of the Subjective in Information Organization and Retrieval," *Journal of Documentation* 43 (September 1987): 200.
22. O'Connor, *Explorations*, 4-6, 21-33.
23. Ibid., 31-35.
24. Ibid., 55, 61.
25. Ibid., 31.
26. James D. Anderson, "Indexing and Classification: File Organization and Display for Information Retrieval," in *Indexing: The State of Our Knowledge and the State of Our Ignorance,* ed. Bella Hass Weinberg (Medford, N.J.: Learned Information, Inc., 1989), 71-72.
27. Ben-Ami Lipetz, "The Usefulness of Indexes," in *Indexing,* ed. Weinberg, 112-113.
28. Ibid., 114-115.
29. Harold Borko and C. L. Bernier, *Indexing Concepts and Methods* (New York:

Academic Press, 1978), 8; and H. H. Wellisch, *Indexing From A to Z* (New York: H.W. Wilson, 1991), xxiii.
30. Harold Borko, "Toward a Theory of Indexing," *Information Processing & Management* 13 (1977): 355, 365; and Dagobert Soergel, "Indexing and Retrieval Performance: The Logical Evidence," *Journal of the American Society for Information Science* 45 (September 1994): 590.
31. Soergel, "Indexing," 591-594.
32. Gerard Salton, "The Smart Environment for Retrieval System Evaluation-Advantages and Problem Areas," in *Information Retrieval Experiment,* ed. Karen Sparck Jones (London: Butterworths, 1981), 318.
33. Soergel, "Indexing," 592.
34. Ibid., 593.
35. Ibid., 593-594.
36. Ibid., 598.

8

Representation Illustrated

The SMART System

From the point of view of the physical metaphor, the uncertainty and shortcomings that accompany the human element of representing the aboutness of texts are problems that need to be and can be overcome. The main step taken to accomplish this goal is automatic indexing. In contrast to human indexing, usually referred to as manual indexing, simple automatic indexing uses computer technology to select potential index terms from a collection of texts by counting the words and ranking them by frequency of appearance. Index terms are then identified and assigned on the basis of their sufficient yet not excessively frequent appearance in the texts from which they are taken.[1]

One of the most powerful conclusions of the Cranfield information retrieval experiments was that single terms, automatically selected from the natural language used in texts and with minimal control for synonyms and variant word forms, resulted in recall superior to that which can be achieved through the use of manual indexing. Some gain in precision can also be achieved from the use of simple term coordination in Boolean searches.[2] This result led to the posing of a question. Is it possible to take advantage of the physical characteristics of language in texts, as explored by Zipf and revealed by the Cranfield experiments, to move beyond the constraints to information retrieval set by controlled vocabularies, manual indexing, and exact match search requirements?

Gerard Salton believed the answer was yes, and he set out to design a retrieval system that illustrated his point.

The classic concerns of research into statistical and probabilistic information retrieval, including automatic indexing and searching, are represented in Salton's SMART (System for the Mechanical Analysis and Retrieval of Text) experiments. These experiments explore the possibility of overcoming the perceived shortcomings of manual indexing by exploiting the nature of language itself. Four techniques, all derived from linguistics, are central to this effort. Each is based on the characteristics of language that allow for regular, predictable, and logical ways of relating words and their meanings, and manipulating the inherent aboutness of texts. As it turns out, not all of these techniques prove equally useful, but the logic and purpose of each is worth consideration. However, the SMART system can use preconstructed hierarchical term arrangements to relate words in a subject area by their level of specificity, thus allowing the focus of a search to be expanded or narrowed. It can use a thesaurus to link concepts at the same level of specificity in order to tease out the nuanced meanings of texts, and it can also employ automatic syntactic and semantic analysis. The former identifies and specifies the syntactic role of words in texts, so that complex content descriptions can be formed while avoiding the kind of syntactical confusion we observed with the boolean search for "venetian and blind." The latter involves identifying the semantic roles played by words and phrases that describe content in way that allows links between concepts that might be missed by simple thesaurus connections.[3]

Like all automatic indexing, what is actually occurs is a kind of content analysis of the texts in a collection and the queries posed to them in order to link queries and texts as an outcome of the analysis. Matching the aboutness of queries and text still occurs, but because of the way concepts are related to words and words can be related to other words through the use of various linguistic manipulations, the match need not be exact to result in the retrieval of texts that are relevant to queries. Even more important is how SMART relates texts to one another on the basis of a composite of the words that represent them. With conventional automatic indexing, texts are represented by a frequency ranking. The more frequently a word appears in a text and the more heavily it is weighted, the more important it is to the aboutness of that text. SMART goes beyond such simple counts. Each text is represented not by a single term, or even a set of discrete terms, but by a vector of terms. In addition, the weight, or importance of a term as an indexing device is not a matter of raw frequency of its appearance in a text but of its ability to separate and uniquely identify the aboutness of texts in a collection. In short, it is a matter of its discrimination value.[4]

Words that occur very frequently in a text or collection of texts are too common to discriminate well between texts on the basis of aboutness. As retrieval terms they will group too many texts together and can compromise precision. Words that occur too infrequently are too rare to discriminate well. They will group too few texts together and do harm to recall. Thus, the ideal term discrimination value is an outcome of an intermediate frequency of occurrence of a word in a text or collection of texts. It is a word that can identify important differences in the aboutness of texts and discriminate between them. The power of this approach to controlling information lies in its recognition and acceptance that texts are dissimilar and that certain words in texts will best reveal this dissimilarity and the best indexing increases the average dissimilarity of texts in a collection.[5] As contrary as it may sound, our ability to get what we want from a collection of texts to maximize our recall and precision at the same time depends not on being able to determine how much texts are alike but rather on how different they are.

In the SMART system, words that occur with a high frequency in a collection of texts are passed through the linguistic mechanisms described above and transformed into terms or term phrases with a lower frequency of appearance. Words that occur with a low frequency are passed through the same mechanisms and transformed into terms or term phrases with a higher frequency of appearance. The goal is to identify those terms that discriminate between texts by revealing their dissimilarity. In the end, the indexing term created may be a single word, a phrase, or string of words; and given the process by which it is created, it may not actually appear in any text in the collection. Even so, the created term may, in fact, better represent the aboutness of a text than any word or phrase actually in the text.[6]

To a point, then, SMART works like any other automatic indexing system, including most Internet search engines.[7] A computer scans every text in a collection, creates lists of every word in that collection, and keeps track of their frequency of occurrence both in each text and in the collection as a whole. Stopwords are then removed. These are words that are necessary to make language work but convey no subject content, such as articles and conjunctions. Once the stopwords are taken out of play, the remaining words—mostly nouns and adjectives—are reduced to wordstems in order to account for all possible variations of use and meaning a word might connote. For example, *library*, *librarian*, and *librarianship* are all reduced to the truncated term *librar*. At this point, the working vocabulary of the collection of texts under treatment is passed through the linguistic transformation mechanisms in order to maximize the term discrimination value of each index term. Next, SMART takes the step of creating word phrases from pairs of words that are within four places of one another in any given text in order to create descriptors for texts. Finally, based on how frequently an index term or phrase appears in a text, term weights are assigned. What follows, however, is what makes SMART truly a unique and very powerful information retrieval system.

Because the physical metaphor of information as a theoretical object allows us to treat texts and words as objects and computer capability allows us to manipulate very large numbers of data objects, idea and technology can be combined in a most interesting way for the purpose of retrieving information. A virtual, multi-dimensional space representing a collection of texts can be defined by a computer that creates an axis for every index term as described by the automatic process above. In other words, this space is defined and created by as many axes as there are index terms and is known as document vector space.[8] Each text in the collection, then, is represented by a document vector whose place in the document vector space is determined by the relationship between that text and the axes that define the document vector space. Figure 8.1 illustrates this idea in a simple two dimensional context.

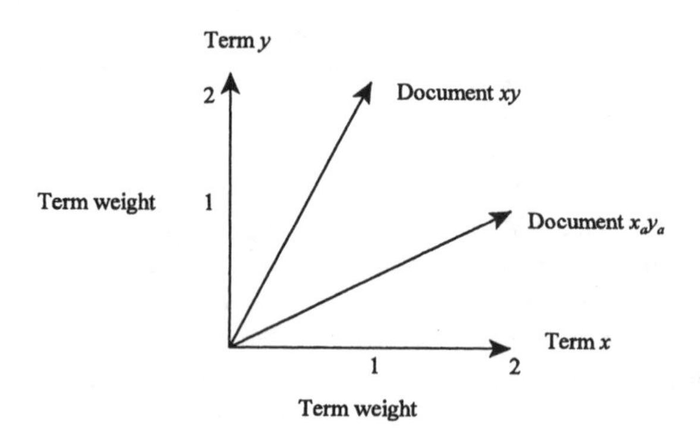

Figure 8.1 Document Vector Space: Two Documents with Two Terms

We begin creating our document vector space with a single text, and based on their frequency of occurrence and appropriate transformation, we identify the aboutness of this text with two terms. Term x occurs with twice the frequency of term y, so by assigning a term weight of 1 to term x, and a term weight of 2 to term y, we can create the vector xy that represents the document in the vector space. If we then add a second document to our collection, again identified by the same two terms, except that now y occurs twice as often as x, we can represent that document with a different vector, $x_a\ y_a$. Notice that we now have two documents that conventional or manual indexing might have represented with the same terms taken from a controlled vocabulary because, given their content, they are about the same subject. In our document vector space, however, each of these documents occupies a discrete position; and their dissimilarity is made explicit by the measurable size of spatial gap that separates the two document vectors.

Now suppose we add to our collection a third document also represented by the same two terms, except in this instance word x occurs in the text 1 3/4 times more often than does word y. In Figure 8.2, this third document is represented by the vector $x_b y_b$. This figure reveals a great deal about these three documents that is useful for the purpose of retrieving them from the collection.

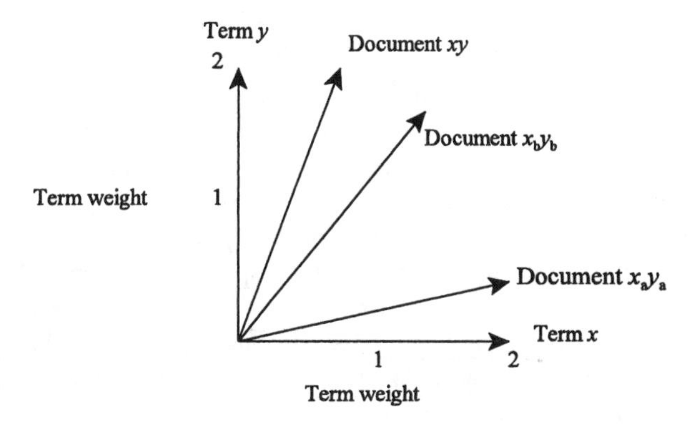

Figure 8.2 Document Vector Space: Three Documents with Two Terms

It tells us that document $x_b y_b$ is less dissimilar to documents *xy* and $x_a y_a$ than either of these documents are to each other, and that document $x_a y_a$ and document $x_b y_b$ are separated by a greater degree of difference than are documents *xy* and $x_b y_b$. Now if we are uncertain of what we really want, except that we know we want documents about *x* and *y*, we can ask the database to yield every document that contains these two terms. In this case, the outcome is identical to searching the database for the match to the search "*x* and *y*." If, however, we know we are more interested in *x* than *y*, we might specify a search that identifies *xy* as the most wanted document and then demands that the extent of space, or dissimilarity, that separates any other document of possible interest must be small enough that document $x_b y_b$ is included in the search results, but document $x_a y_a$ is not, because it is literally too far away in aboutness. The difference between its content and that of document *xy*, as quantified and represented by the space between their vectors, is too great to make it of relevant interest. Remember, that in addition to making this distinction for us, the system may also have transformed the words we used in our query into the terms *x* and *y*. Thus, the final outcome is the retrieval of documents relevant to our query with high levels of recall *and* precision, all accomplished without a single exact match between any word we used to pose the query and any word in any retrieved document.

A vector relationship can be established between each text in a collection and the contents of the entire collection as represented by the document vector space. Each document is represented by a vector that occupies a unique physical location in the *n*-dimensional space created and defined by all the important terms and concepts present in and expressed by the entire collection. This space, then, has as many dimensions as there are such concepts and their representing terms. It may at first seem that for a collection of texts of any reasonable size and the number of different content-bearing words they contain, the number of dimensions of the collection document vector space will be impossibly large; but keep in mind the essence of Zipf's law. Given the economy of language at play by the principle of least effort and the action of reducing the number of terms at play by deliberately transforming them into terms of intermediate frequency of occurrence, there may not be as many term-dimensions defining a document vector space as one might first expect.

Within a given a query, words can be subjected to the same kind of content analysis as is conducted on the texts of a collection. From such an analysis, a query vector can be constructed and placed into the document vector space. Documents can then be retrieved and ranked by the probability of their relevance to the content of that query on the basis of the physical proximity document vectors have to the query vectors in the document vector space. In other words, the retrieval of a stored item in response to a query depends on the magnitude of a global similarity coefficient measured by the size of the angle between a query vector and a document vector. The threshold of dissimilarity beyond which a document will not be retrieved can be specified exactly and quantitatively in terms of the cosine of angles between vectors. Beyond a specified cosine, then, documents are not retrieved. As well, the cosine of the angle between the query vector and the document vectors is a measure of the probability that a document will be relevant to the query. Thus, the smaller the cosine of the angle between the vectors, the more likely the retrieved document will be relevant. If users of the system decide that a search produces insufficient recall, they need only increase the magnitude of the acceptable cosine between query and document vectors. If precision is thought to be lacking, the cosine can be decreased.[9]

Since entire vectors are compared, the basis for document retrieval is a composite similarity between query and document vectors where the document vector space is constructed so as to discriminate between documents, that is, to place them, based on the difference in their content, as far from each other as is possible in the document vector space. As a result, an exact match between query *terms* and document *terms* is not required. In addition, since the measure of global similarity is quantitative, the threshold for retrieval can be set to cast an exactly specified wide or narrow net in a manner that is completely controlled and predictable. System output can then be ranked by the magnitude of document similarity to a query and presented in a ranked relevance order. And all can be accomplished without resort to Boolean search logic and its ambiguities.

An expansion of the examples presented in Figures 8.1 and 8.2 is presented by Figure 8.3 to illustrate SMART at work. This figure represents a multi-dimensional document vector space of three dimensions, where each dimension represents a term created from a concept taken from a document in the collection, and transformed into a representative term. Three documents from the collection are represented by vectors constructed from the relative weights of each term they contain. Document *xy* is represented by only terms *x* and *y*; document *yz* is represented by only terms *y* and *z*, and document *xyz* is represented by all three terms. Remember to keep in mind that given two documents, the size of the angle between their vectors might begin as a very large one based on their difference regarding the weight of the first term. The effect of the second term, however, might be sufficient to bring their vectors back together by a closer matching of the weight of that term for each document. The query vector (*qx,y,z*) is then placed into this document vector space. Notice that only document *xyz* contains exactly the same terms as the query vector and that documents *xy* and *xyz* are less dissimilar to one another than either is to document *yz*. It is easy to imagine a situation in which the similarity threshold is set to retrieve only documents *xy* and *xyz*, that is, a relatively small angle between query and document vectors is specified in order to maximize the recall and precision of the search.

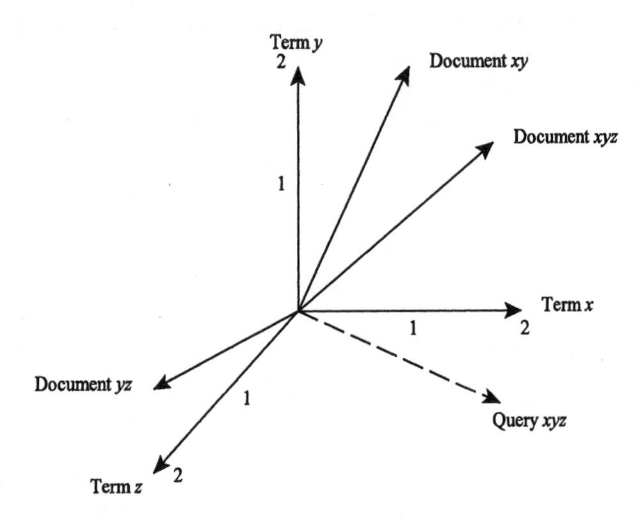

Figure 8.3 Document Vector Space: Three Documents and Query

Various experiments have proven Salton's SMART system to be a remarkably powerful indexing, searching, and retrieval mechanism. It turns out that overspecification of document aboutness can be just as bad as underspecification for the purpose of retrieval. Of the various linguistic mechanisms used to transform the aboutness representation of documents and queries, syntactic analysis and concept hierarchies appear to be of limited usefulness. The extraction of weighted word stems from titles and document abstracts, however, supplemented with the use of a classification system or thesaurus to recognize and take advantage of synonyms and related terms, appear to be the most useful mechanisms for creating effective document vector spaces. When it comes to designing information retrieval systems based on the physical metaphor, a certain complexity must be accepted; but within that context, the simpler is clearly the better.

The vector space model does many things exceptionally well, among them the ability to make intertextual relations explicit without recourse to human interpretation. It makes possible a kind of indexing that reflects document content and distinguishes between documents on the basis of their decreasing similarity. This approach to bibliographic control relies on a process of reducing the meaning of documents to the status of an object, in this case, a document vector, but it is a powerful reduction. It allows documents and, indirectly, their meaning, to be manipulated and compared in a manner that provides a measurable degree of quantitative certainty regarding the interpretation and representation of documents and their relevance to search queries. The problem of representation is effectively reduced to a technological problem free from the subjective influences and ambiguities that characterize human interpretation of the subjects of texts.

Salton's SMART system reveals the value of not allowing information to be something other than its representations. By imposing this limit and treating information merely as an object, we can render it considerably more predictable and controllable. Of course, to do so we must accept the possibility that we may not really have a grasp on what it is we are successfully doing. Like all approaches to information based on the physical metaphor depend, the approach upon which SMART depends relies on a number of critical assumptions. First, and arguably the most problematic, is the notion that texts explicitly communicate their aboutness in ways that can be objectively assessed. Second, the assessment of aboutness can take the form of a quantitative measure constructed from word counts and frequency distributions. Third, word meanings are generally reliable, constant, and can be accepted at face value. Finally, the context created by other words within which a word appears is not problematic for the meaning of that word. If these last two assumptions, because of some unlikely and unexpected turn of chance, are problematic, then linguistic analysis of texts can be applied to solve the problem.

The ASK Model

There is another approach to the control, retrieval, and use of information, however, that does not share these assumptions. From the perspective of the cognitive metaphor, the variability of meaning associated with language makes all attempts to definitively determine and predictably manipulate the aboutness of queries and texts an uncertain enterprise. By limiting relevance to a matter of matching the aboutness of queries to the aboutness of texts, even if an exact match is not required, physical metaphor-based solutions diminish the reason someone uses an information system in the first place. The query substitutes for the problem, and the problem is reduced to merely a query. Remember that from the cognitive perspective, people do not use information systems to retrieve texts but to retrieve texts in order to use their contents to solve problems. In order to claim that an information system is effective, then, it must provide information that makes a difference to the user. The outcome of its use must be a change of mind. Solving problems of aboutness by reducing texts and queries to their representations is not helpful unless we can solve the problem of what users are about when they seek out information.

To this end, Nicholas Belkin, Robert Oddy, and Helen Brookes set about designing a prototype information retrieval system that challenged a number of traditional assumptions regarding conventional indexing and automatic alternatives. They begin from the assumption that information needs result from inadequate states of knowledge, also known as an anomalous state of knowledge (ASK). Within the context of the cognitive metaphor, the relationship of query to need is not necessarily straightforward. A query may be perfectly represented to a system of information retrieval, but the query itself may be a less than perfect representation of the problem troubling the information seeker who poses it. In order to solve users' problems, then, their needs must be represented in terms appropriate to the actual task they must accomplish. Belkin, Oddy, and Brooks argue that in order to be truly effective, information retrieval systems must be constructed in such a way as to account for and represent a user's ASK to the system. While the representation of queries and texts is still important, a more important issue involves recognizing that users are generally unable to specify precisely what they need, even when they are quite descriptive about their problems. Representing users and their problems to the system is more important than representing texts to users.[10]

The cognitive metaphor's approach to representation challenges two central assumptions on which the physical metaphor depends, that it is possible for users to specify their information need as a discrete and precise query and that information needs, by means of this specification, are made functionally equivalent to the texts that will answer a query. By insisting that only those concepts and their relations that are explicitly stated by a user are significant to retrieval, the physical metaphor divorces information needs from the problem situation that is their source.[11]

In contrast, the cognitive metaphor, by assuming that information is only a means to an end, recognizes that seeking information is a sign that an individual's current state of knowledge is inadequate to the achievement of a desired end. An anomaly in the state of knowledge stands between a desired end and its realization, and one must affirmatively decide that information might address the anomaly before it is even possible to begin a search for information. One must transcend a visceral condition in order to transform it into a conscious need capable of being articulated. This assertion supposes that even a scientist contemplating a problem regarding the nature of the universe, let alone a suburbanite pondering how to build a doghouse, does not immediately conceive this problem in terms of a set of texts that needs to be read in order to solve the problem. To the scientist, the problem is one of nature; to the suburbanite, one of how to house his dog. It is not likely that either of these people immediately perceives their problem in terms of information and its retrieval.

On the contrary, it is very likely that both of them, when at an early stage in the process of thinking about the problem and regardless of what or whether knowledge can be brought to bear, will be unable to fully articulate exactly what information might be of help. Later, when the time comes to design an experiment to demonstrate a hypothesized principle, or actually build a doghouse, texts describing similar experiments or examples of doghouses might be useful. Queries and documents, as texts, represent fundamentally different phenomena. A document is a statement of something known. A query is a statement of something unknown, and so we would be foolish to assume that these two phenomena are really very much alike.

Associative retrieval and relevance feedback, manifest in information retrieval systems based on the physical metaphor, attempt to resolve this dilemma. This approach has achieved some success by extending the reach of a search for information without an knowledge or representation of users' needs, but it does not explicitly address the issue or existence of an anomalous state of knowledge or a user's inability to actually articulate one. Belkin, Oddy, and Brooks argue that the progress made toward the satisfaction of needs by such an approach is at best marginal, because the doubt, uncertainty, and suspicion of inadequacy that characterize a real information seeker's condition are not allowed to be at play. They suggest the alternative of attempting to represent the anomalies caused by and associated with information-seekers' large-scale intentions and goals. Thus, their approach neither demands or expects that information-seekers will be able to specify the information they need in advance of its retrieval and recognition.

A successful information retrieval system will be one that begins with the recognition that it must be regarded as a user-initiated system of communication and that the driving force of an information-seeking situation is the user's problem. Any resolution of a user's anomaly that might be provided by retrieved information must be evaluated in terms of this problem and the user's decision regarding whether or not the user's view of the problem changed as an outcome of using that information. The fundamental representational issue at stake is a matter of building a model of the user's aboutness, including whatever anomalous state of knowledge is a part of that and, then, identifying a text whose aboutness addresses that model. The representation of the user is the point from which the representation of texts begins, and this determines the mode of representation that will be imposed on a collection of texts. In other words, the system must be made to adapt to the person, rather than the other way around.

To demonstrate the possibility that such a retrieval system would work, Belkin, Oddy, and Brooks designed an experiment. While many developments along these lines have since taken place and considerably more sophistication has been achieved, this early work still reveals the basic principles at play. To solve the problem is one of representing the structures of a user's information need and the texts that might satisfy that need, the ASK system relies on mechanisms similar to those of the Cranfield experiments and SMART. In this respect, its method of representing texts and queries that is not all that different from an approach based on the physical metaphor of information as a theoretical object. Even when attempting to cope with information as a cognitive phenomenon, it is very difficult to avoid or transcend its material attributes.

First, a problem statement is elicited from a user. This statement is supposed to be unstructured, about two or three paragraphs in length, and intended to present the user's problem rather than pose a specific query or request for information. Then a model of the user's anomalous state of knowledge is constructed by means of an automatic content analysis of the problem statement similar to those employed in all automatic indexing. It is based on the assumption that the physical proximity of words to one another in the problem statement can be taken as a sign of the conceptual relations that inform, perhaps confuse, but in any case contribute to a user's state of knowledge. The model consists of a structural representation of the strength of association between every pair of significant words in the problem statement.

If the two words appear in the same sentence and are adjacent to one another, the conceptual association is taken to be a strong one; if they appear in the same sentence but are not adjacent, then the association is taken to be of a medium level; and if they appear in adjacent sentences in the same paragraph, the association is taken to be a weak one. The outcome of this content analysis is a conceptual map of the user's state of knowledge that relates all of the present and identified concepts at play according to the quantitative strength of their association. The user is then asked to examine this map, and judge the adequacy of its representation. This relevance feedback can then be used to adjust the model, even before a search takes place.

A crucial assumption here is that truly useful information search outcomes depend on posing queries that best represent a user's problem and the needs associated with that problem. The goal is to help the user ask a better question.[12] The model of the user's anomalous state of knowledge can then be classified, essentially by a measure of the extent to which the concepts comprising it are coherently integrated and related. A model revealing only a few points of connection between concepts strongly associated with one another represents both a well-defined problem and a user who is already well on the way to understanding and resolving his or her anomaly. Conversely, a model revealing a number of connecting points between concepts that are only weakly related represents a user whose problem remains largely undefined.[13] The latter user might be prompted to rethink his or her problem.

Next, the same kind of content analysis is applied to the texts in a collection, and the model of the user's state of knowledge is mapped onto the model of the text's state of knowledge. The comparison of these physical representations of states of knowledge reveals the influence of the physical metaphor. Nevertheless, retrieval of a text is based on how well the entire structure of the former matches the entire structure of the latter, such that an exact match between particular concepts, associations, or their relative strengths is not necessary to retrieval. Abstracts, rather than full texts, along with a rationale for their selection, are presented to the user, in order to initiate a dialog between system and user. From the user's responses regarding the relevance of the abstracts, the system infers the user's evaluation of the search strategy, the suitability of the text to the problem, and the need to reinterrogate the collection in order to improve retrieval results. Note that this process can result in a wholesale rethinking and restatement of the original problem. The information-seeking process is not limited to a single discrete event but is instead conceived as a series of iterations, repeated until the user is satisfied with the outcome.[14]

The MONSTRAT Model

Early tests with the ASK model demonstrated that it is possible to obtain user statements that represent anomalous states of knowledge. Content analysis of full-text is entirely possible for the purpose of building models of the knowledge structure of texts, but abstracts apparently serve as sufficient document surrogates. It was clear to Belkin, Oddy, and Brooks that their method of textual analysis and representation resulted in oversimplified models that tended to overlook the importance of weaker conceptual associations and that their approach to classifying ASKs was unrefined. Overall, however, they concluded that the model worked, and that their theory regarding the existence and role of anomalous states of knowledge in information representation and retrieval had been confirmed.[15]

The most important outcome of this experiment is that, despite its reliance on the physical metaphor, it transcends that metaphor by suggesting that information retrieval must be about something more than the effective manipulation of texts and their representations. The user and the user's cognitive characteristics must also be a part of the system, which in turn must be conceived as an interaction, perhaps through representations of texts and users, but ultimately between texts and users. In short, it must be a system of communication.

This notion is further illustrated by the MONSTRAT (MOdular functions based on Natural information processes for STRATegic problem treatments) model that was developed from the ASK experiments. The MONSTRAT model specifies what an information retrieval mechanism must actually do to help a user solve a problem. Thus, an information seeker's situation is at the heart of the MONSTRAT model. To even begin a search for information, then, certain aspects of this situation must first be determined and represented for use by an information retrieval system.[16]

The fundamental assumption underlying the MONSTRAT model is that an information system of any kind must apprehend and model the information problem that brought the user to an information retrieval system in the first place.[17] The model itself derives from the observation that successful information retrieval, regardless of how success is defined, depends on an interactive process of model-building that engages a user and some kind of intermediary, possibly a librarian, representing the information retrieval system. Together they construct both a model of the user's state of knowledge and the problem that conditions and gives rise to an information need.

This interactive information retrieval process is constituted by certain specific functions that must be successfully performed in order to accomplish the task of helping a user to solve his or her problem. An information retrieval system must acquire its knowledge of the user by means of a dialog with the user and, then, apply this knowledge to the search and retrieval process. None of this is unfamiliar to a reference librarian. Everyday, reference librarians conduct interviews with users of libraries. Some of these exchanges are quite brief, others are extensive, but in any case, the librarian acts as an intermediary and represents the library and its information resources. By means of a dialog with the user, the librarian arrives at an understanding of the user's need in order to select the appropriate sources, conduct appropriate searches, and retrieve information relevant to the user's need. The MONSTRAT model was created to capture the reality of this human process and use it to design supportive and "intelligent" automated information retrieval systems that could function in place of the human intermediary.[18]

To accomplish this end, the MONSTRAT model specifies ten functions that human-machine interaction must perform, for which techniques were developed to accomplish each one.[19] The first function that must be performed is to determine the appropriate *dialog mode* between the user and the system. The basic choice is between the use of natural language, as in the early ASK experiments, and the use of a menu-driven mode, and is essentially a matter of how the user and the system will talk to each other. Already we can see that the MONSTRAT model is positing a system that will work behind an interface with the user, that this work will be invisible to the user, and that what the user ultimately sees is the results of this work. MONSTRAT-based retrieval systems work with a form of automatic discourse analysis, potentially based on a number of linguistic analysis techniques employed by the SMART system, in order to build models of and match the states of knowledge presented by users and represented by texts. As we shall see, while such a system must also present to its user an explanation of how it does its work, it does not reveal this work in action. In this respect, it is no different from either a SMART system or the thinking processes of a reference librarian.

The initial user input to a MONSTRAT-based system is neither a query nor a search statement but a problem statement, presented in a form determined by the performance of the dialog mode function. The system uses this input to fulfill its second necessary function of determining the user's *problem state*. This function identifies the dimensions of the problem, establishing the context of a user's ASK, and includes an accounting of a user's information-seeking activities regarding his or her problem prior to engaging the information retrieval system. The third function that must be performed is establishing the *problem mode*. The purpose of this function is to explain the capabilities of the system to the user, and to clarify how the system works. At this a point, the user should be asked to decide to use an alternative means of resolving his or her problem, and the ideal system will be capable of providing referrals to appropriate alternatives. The fourth function requires the system to build on its knowledge of the problem state in order to construct a *user model* that characterizes for the system its user's goals, status, and familiarity with the subject of the problem, as well as with information retrieval processes in general. Closely related is the fifth function of *problem description,* which requires the system to build a model of its user's specific problem. The essential elements of this model include the user's topic and task, his or her particular approach to the topic and task, the content of the texts that the user desires, and the background knowledge necessary to provide a context for any retrieved information.

The problem statement, user model, and problem description, taken together represent the knowledge of the user that the system/intermediary must have in order to build a model of its user's ASK and initiate a search for information. However, we must remember that the model of the user and his or her information problem that the system builds is automatically analyzed from statements made by the user to the system. During this process, concepts related to the user's need are identified, related, and associated with other concepts. Syntactic analysis can be applied to resolve "Venetian blind" problems, plus a variety of other methods are available to measure the intimacy of association among the concepts that appear to delineate the user's problem. Finally, we must also bear in mind that the user and the system are engaged in a dialog. We should not assume that the process of building a model of the user and his or her problem is linear, but that the system should provide feedback within the context of every function, allowing the user to modify his or her input, and in turn the system to revise its model of the user and the problem. In effect, the MONSTRAT model provides for a kind of relevance feedback as an outcome of user/machine dialog before the retrieval of any information.[20]

Still, the five functions we have examined so far must be performed before it is possible to move to the next three. The sixth function requires the development of a *retrieval strategy*, including the selection of databases, the selection of search terms, and the formulation of a search query appropriate to the user/problem model. The seventh function requires the employment of a *response generator* that is used to determine and customize the propositional structure of the system's responses when engaged in dialog with the user. At this point, the user selects from among alternatives how she will "talk" to the system. The eighth function is accomplished by the *input analyst*. The execution of this function automatically converts input from the user generated from the user/system dialog into structures that can be used by search mechanisms to explore the database. This process essentially is a matter of converting the statements representing the user/problem model into physical structures in a manner similar to that of the original ASK experiment, thus allowing the structure of a user's anomalous state of knowledge to be mapped on to the knowledge structures of the texts in a database. All three functions are clearly central to the actual search and retrieval process of a MONSTRAT-based system. In each case, while the system's actual work will be invisible to its user, it is nonetheless represented to that user, and further user input, revisions, and corrections will be solicited.

The last two functions of the MONSTRAT model are about delivering and explaining information retrieval system output. The ninth *output generator* function converts the system's propositional output into a form appropriate to the user and the contextual variables that define his or her situation. This action involves the presentation of search results, display options, and menu options that allow a user to select, order, examine, and manipulate the results. The outcome of any of these actions can then serve as an input to the system, causing it to repeat one or more of the MONSTRAT functions described here in order to revise and refine the results of its search. The output generator function creates yet another point at which iterative possibilities are built into the system. The tenth, and final function is *explanation*, which explicitly reveals the operations of the system to the user. Its purpose is to provide a description of how and why particular output was selected and presented in a particular form in order to allow the user an opportunity to exercise his or her own judgment regarding the adequacy of the system's work.

Since its inception, the MONSTRAT model of information retrieval has generated some controversy.[21] While many information scientists welcomed its focus on the user, others noted its exclusion of a means for revealing to a user the cognitive structures and assumptions built into and employed by the system, and in so doing provide that user with an additional system control. The MONSTRAT model also does not include a domain knowledge model of the concepts and relations between concepts that comprise the subject knowledge domain inherent in the texts and scientific literatures to be searched. A domain knowledge model could allow for the analysis of user input in terms of the knowledge available and provide a means for comparing user knowledge and subject knowledge domains, especially regarding appropriate paradigmatic structures. Despite these criticisms, there is widespread agreement that the MONSTRAT model successfully identifies the functions that must be performed by information retrieval systems, and that it provides a foundation for further development of "intelligent" intermediary systems.

The MONSTRAT approach is represents a considerable advance beyond the ASK model. The research on which it is based focuses on both the nature of user/system interaction and the question of how to integrate this knowledge into working information retrieval systems. In turn, a number different experimental information retrieval systems have been designed around its principles, all of which display two fundamental features. First, each constructs a model of the user in the system, and then derives from this model the cognitive characteristics of that user.[22] In this context, this condition is a prerequisite of information retrieval. Second, in each case the user does not interact directly with the database, but through a representation of his or her perceptions, requirements, preferences, and state of knowledge. In principle what is represented and searched for are conceptual associations; in practice these associations are manifest as word associations substituting for the knowledge structures of texts and users.[23] Ultimately these models attempt to make the user a part of the system in such a way that the representation of texts derives from a representation of their potential user. The underlying notion is that the means to transcend the constraints and limits of the physical metaphor, which is still regarded as necessary but not sufficiently theoretical, can be achieved by applying a cognitive metaphor to the problems of information representation and retrieval.

There is little doubt that the MONSTRAT model and the cognitive metaphor that informs it effectively captures the complexities of dealing with information. Together they raise a variety of fascinating questions, and they suggest the possibility of designing new and even more powerful systems. While it seems plausible that some combination of technologies based on the paradigmatic principles illustrated by SMART and MONSTRAT could lead to the development of practical user-driven, intelligent information retrieval systems, this result has not yet come to pass. Most efforts along these lines remain little more than experiments carried out with test collections and queries in computer laboratories.

At the same time as the MONSTRAT model captures complexity, it reveals the shortcomings of the cognitive metaphor. As suggested earlier, this metaphor does identify a puzzle to be solved but it is difficult to combine its pieces into a coherent theoretical or practical whole. Work continues along these lines, but expert systems information retrieval remains elusive, especially for non-specialized users and knowledge domains. Prototype systems still employ relatively simple search experts and domain knowledge bases, and useful evaluation is difficult.[24] The interactive and flexible nature of most search sessions makes it difficult to establish standard expectations as well as compare one search against another.

Conclusion

The strength of expert systems is also their weakness. Because they allow such customized responses to individuals, it is not easy to gauge their overall utility for a number of different users. They accurately reflect the complexity of the information retrieval problems in their number and variety of parts, but because there are so many working parts, including model builders, search mechanisms, interfaces, and explainers, it can be very difficult to sort out which parts are working well and which are not.

Expert systems certainly provide more assistance and explanation to their users than do conventional information retrieval systems, but here again, a trade-off is involved. By making more choices and providing more freedom for their users, these systems also run the risk of causing more confusion. They demand considerably more engagement from their users than do conventional systems, and they depend heavily on the user's willingness to put forth the effort required to get them to work.[25] There may also be a point of diminishing returns for human beings regarding information-seeking behavior.

For while it may be possible to invent a system that comes close to perfect retrieval, for most people such perfection may not be required. Most information problems, because of their indeterminate nature, may instead require a series of user/system encounters before their users can arrive at a true understanding of their problem, regardless of the system used to find information. Reflecting both the promise and the problems of complexity inherent in the cognitive metaphor's notion of information as a theoretical object and the efforts to create expert systems based on that notion is a lack of a fully developed idea of what an expert system is and what it is out to do. At least one of the original participants in the original ASK systems experiments argues that the difficulties of designing working expert systems have been underestimated.[26] Information retrieval does not lend itself easily to expert system solutions, not because it is not well defined or well understood, but because it can be defined and understood in so many different ways. In short, finding the necessary expertise to import into the system is difficult because there is, as of yet, too much that not known about the nature of information, its organization for access, and its users.

Similarly, SMART and MONSTRAT are interesting not because they have led to the development of practical application, but because they are representative of ideal type information retrieval systems that embody the unity of opposites manifest in the physical and cognitive metaphors. Despite their obvious differences, they display a number of shared characteristics. Both systems employ integrated "experts" of different kinds to cope with problems of semantic interpretation, and each actually executes the tasks of the other, albeit with a different emphasis.

SMART's document vector space and the document vectors plotted within it originate in the content and meaning of texts. Control over and access to information is a quantitative measure of metaphysical space, defined by a virtual reality with material spatial dimensions. The amount of virtual space between the vectors represents the content differences between documents, allowing them to be identified and related for the purpose of retrieval. Judgment and interpretation are then transformed into a phenomenon that can be apprehended in a language of physical things. Cognitive freedom to interpret the results of access is present in feedback mechanisms in which both words and measures of *closer* and *farther* can be used to reconfigure and specify system output.

The ASK and MONSTRAT experiments begin with an explicit recognition of the kind of usually incompletely defined need that motivates a search for information and the implications of this condition for information retrieval. The models they build are physical representations of the content of a user's mind, or at least a user's problem state, on one hand, and the content of documents on the other. While the content of cognition and its actual use in resolving a problem is of primary concern, these systems cannot accomplish their tasks until these cognitive phenomenon are transformed into matching physical models. In addition, the communication these systems seek to facilitate between a writer and reader must be mediated by a text representing the writer, who is neither present nor active in this process of communication.

Ultimately the access to information created by SMART, ASK, and

MONSTRAT depends on a phenomenon identified earlier—the encounter between an informative object and an interpreting subject. The implications of this encounter for the phenomenon of relevance is the subject of the next chapter.

Endnotes

1. Gerard Salton, "The SMART Environment for Retrieval System Evaluation-Advantages and Problem Areas," in *Information Retrieval Experiment,* ed. Karen Sparck Jones (London: Butterworths, 1981), 317.
2. Karen Sparck Jones, "The Cranfield Tests," in *Information Retrieval Experiment* ed. Karen Sparck Jones (London: Butterworths, 1981), 278.
3. Salton, "SMART Environment," 318.
4. Ibid., 324.
5. David Ellis, *New Horizons in Information Retrieval* (London: Library Association, 1990), 27-28.
6. Salton, "The SMART Environment," 325.
7. Susan Maze, David Moxley and Donna J. Smith, *Authoritative Guide to Web Search Engines* (New York: Neal-Schuman, 1997), 21-31.
8. Ellis, *New Horizons*, 29-30.
9. Salton, "SMART Environment," 319.
10. N. J. Belkin, R. N. Oddy and H. M. Brooks, "ASK for Information Retrieval: Part I. Background and Theory," *Journal of Documentation* 38 (June 1982): 61-71; and N. J. Belkin, R. N. Oddy and H. M. Brooks, "ASK for Information Retrieval: Part II. Results of a Design Study," *Journal of Documentation* 38 (September 1982): 145-164.
11. Gernot Wersig, "Information Science: The Study of Postmodern Knowledge Usage," *Information Processing & Management* 29 (1993): 235-236.
12. Belkin, Oddy and Brooks, "ASK for Information Retrieval: Part II," 147-152.
13. Ibid., 152-158.
14. Ibid., 160-161.
15. Ibid., 159-160.
16. N. J. Belkin, H. M. Brooks and P. J. Daniels, "Knowledge Elicitation Using Discourse Analysis," *International Journal of Man-Machine Studies* 27 (1987): 127-130.
17. Peter Ingwersen, *Information Retrieval Interaction* (London: Taylor Graham, 1992), 198-199.
18. Ibid., 92, 109-110.
19. N. J. Belkin, T. Seeger and G. Wersig, "Distributed Expert Problem Treatment as a Model of Information System Analysis and Design," *Journal of Information Science* 5 (February 1983): 157; and Belkin, Brooks and Daniels, "Knowledge Elicitation," 134.

20. R. N. Oddy, "Information Retrieval Through Man-Machine Dialog," *Journal of Documentation* 33 (March 1977): 1-14.
21. Ingwersen, *Information Retrieval*, 110-112.
22. Ibid., 161-174.
23. Ellis, *New Horizons*, 67-68.
24. W. Bruce Croft and R. H. Thompson, "I^3R: A New Approach to the Design of Document Retrieval Systems," *Journal of the American Society for Information Science* 38 (November 1987): 401-403.
25. Ibid., 402.
26. H. M. Brooks, "Expert Systems and Intelligent Information Retrieval," *Information Processing & Management* 23 (1987): 378-379.

9

Relevance

The concept of relevance is intimately linked to judging the effectiveness, and ultimately the usefulness, of information retrieval systems. As we saw with the Cranfield experiments, the central issue in evaluation involves measuring how well a system retrieves documents relevant to the queries submitted to it. Relevance, in this context, makes three assumptions. The first assumption hold relevance to be a function of how well the topic of a text matches the topic of a query. The second assumption holds that the aboutness of texts and queries, as well as their relation to one another, can be objectively assessed. If the topic of a text is regarded as an invariant characteristic of that text, then it can serve as a practical, observable, and measurable basis for making judgments regarding the relevance of texts to queries. The third assumption holds that generally a query adequately and accurately signifies an information need and that retrieved texts adequately and accurately signify information that will satisfy a need. From this perspective, relevance appears to be a relatively straight forward relation between need and information.

Information is either relevant to the reason I am searching for it or it is not. Information systems either recall texts related to the topic of my query or they don't. If I ask about dogs, for example, and get back texts about building, then those texts are not relevant, and the problem is likely with the retrieval system. If I ask about dogs and get back texts about dog training instead of dog care, the retrieved texts are only partially relevant. They are not exactly about the topic of my query, but I asked an imprecise question and failed to specify what I really needed. Thus, I would have to concede that if I wanted texts about dog care and asked only for texts about dogs, then the retrieval problem is mine and not the system's. In fact, the system's relevance judgment was actually successful within the context I had unwittingly set with my question. However, if I reiterate my search I can provide relevance feedback to the system, narrow and specify the topic of my search, and ask a question more likely to produce results that will satisfy my need.

This approach to information seeking and relevance is so common we tend to accept it as almost natural, but here too, it relies on a number of assumptions. First, it assumes that I know what I need or can easily be led to that knowledge. Second, it supposes that my need for information—my anomaly—is so well known to me that I need only use the system itself rather than the information it retrieves to determine if the results are satisfactory. Finally, it assumes that relevance is exclusively a matter of aboutness. A text is relevant to my need if it is about my query, which is itself assumed to be an adequate and accurate expression of my need for information. Thus, relevance and need are linked through a connection between my query and the information a system returns in response to it. Most functional information retrieval systems are based on this view of relevance, but there is another way of thinking about it.

The ordinary meaning of the concept suggests that to be relevant is simply to be related to a matter in hand. An event is relevant to me because it is related in some way to my life. The connection between the event and my life is not necessarily a relation of aboutness; the event may or may not be about topics that comprise my personal or work interests. Rather, it is relevant to me because it changes my life in some way or at least changes the way I think about my life. If, for example, a text addresses a given topic, then it is about that topic. But if a text allows me to see my world differently, if it reveals to me connections between ideas that I did not previously see, or if it strengthens or weakens one of my beliefs, then it is relevant. Viewed from this perspective, an event, a text, or a person may be relevant to me regardless of what any of them may be about. Consider: if I am seeking information, do I want documents about the subject of their queries, or do I want documents whose use will make a difference to me and the problem that motivated me to seek information in the first place?[1]

This question shifts our attention away from a system view and toward a user-centered view of relevance. By focusing on the destination of communication rather than its sender, it provides a reference point for the idea that relevance is a psychological rather than a textual phenomenon.[2] From a user-centered perspective, relevance is about more than just matching the topics of queries and texts. It also includes judgments by users about the extent and degree to which a text is actually useful, while recognizing that it is entirely possible for texts to be useful in some way and still be completely unrelated to the topic of an original query. A text specifically about a query may have a head start, but the key to relevance lies in the relation between a text and its user.[3] Any evaluation of textual relevance, or for that matter the effectiveness of an information retrieval system, is closely linked to an individual's experience, perceptions, and cognitive state at the moment of use.[4] Because it is grounded in the cognitive metaphor of information as a theoretical object, the user-centered view of relevance depends on the seeker's subjective contexts, personal needs, and problem situation.

The assertion that topicality is an inadequate criterion for judgments regarding the relevance of information to a user implies that a number of the assumptions that sustain the system view of relevance, as well as its implicit model of information seeking, must be rejected as untenable. We cannot assume that subject terms in formal queries always adequately represent an information seeker's need. We should also be very careful about assuming that subject terms assigned to texts to describe their content will do full justice to their aboutness. Finally, we cannot assume accurately matching terms in formal queries with terms assigned to texts by formal means of representation will necessarily results in the retrieval of texts that are relevant to needs. Instead, relevance must be regarded as a construction by the one who uses that information.

Like aboutness, relevance is not inherent in texts, and the evaluation of the effectiveness of any system of information storage and retrieval must be closely linked to the experience, perceptions, and cognitive state of the user of the system. In contrast, a user-centered view of relevance implies that the relevance of an event to one who experiences it is an outcome of many elements; topicality or aboutness is a starting point, but the event and its aboutness are assigned meaning in the context of a personal situation. Not only can the same information be relevant to a greater or lesser extent and in different ways to different people, but the same condition may also hold true for a single person who regards the information at different times. The key to these differences lies in the situations and contexts from which the meaning and relevance of information is constructed by its user. In this view, the text as a fixed object is arguably a central variable, but relevance is now seen as situation specific.

Where all of this inevitably leads is to a reconsideration of information need as a theoretical object. The destination view of relevance supports the notion that people do not go to libraries to get books. They go to get the knowledge that is in books. They go to libraries to interact with the content of books. They go in hope that what the author has to say will change their situation. The system view of relevance, and information retrieval generally, posit that an information need is a discrete, static, characteristic aspect of an information seeker and that it can be represented as a query to the system. At the heart of the user-centered understanding of relevance is the notion that a need for information is an unstable psychological condition. It cannot be neatly summarized and stated as a search query.[5]

The relationship between a text and its user is dynamic because the need motivating the inquiry that creates this relationship is dynamic. The need for information can be represented as an anomalous state of knowledge because, like all needs, it emerges from the anomalies generated by life and the changing situations we experience. The need for information emerges from the problems posed by living. More than a characteristic of information seekers, it is a condition of their existence. Being in need of information is itself a situation. This situation includes all the baggage I typically bring to any situation including prior knowledge and experience, an awareness of possibilities and constraints, expected outcomes, and emotional responses.[6] The relationship between a text and its user is mediated by the condition of need, which in turn is affected by all of these factors, and these factors are themselves open to change.

Information science must assume that reason and intention inform communication, and that in turn relevance as an attribute of information and as a theoretical object itself has something to do with effective communication that satisfies some kind of need.[7] Unfortunately, this assumption tends to define one indeterminate idea by reference to another. Positing that effective communication results from the transmission of relevant information from sender to receiver is not substantially different from claiming that the transmission of relevant information from sender to receiver is the defining characteristic of effective communication. The only conclusion available is that some element of relativism cannot be avoided when thinking about relevance in general and the relevance of information to need in particular. Just as some events in my life are more relevant to me than other events, some information is more relevant to a query or need than is other information. All information about the same topic may in some sense be equal, but from another point of view, some potentially informative objects are more equal than others. The difficulties, complexities, and ambiguities surrounding the issue of relevance and effective communication contribute to the attractiveness of the system view of relevance. While the latter may represent a reduction of reality that does some damage to its subtlety, it also allows us to consider relevance in an understandable and manageable way.

Given the ambiguities associated with conventional approaches to the evaluation of information retrieval systems and the difficulties of assigning aboutness to texts, none of this should be surprising. Once this condition is accepted, however, we can begin to think about and construct a more compete understanding of relevance. The value of doing so is first related to an awareness of how and why information retrieval systems fail to achieve their desired ends and, second, to the variability, complexity, and volatility of the human beings who desire these ends. It is entirely possible for inadequate and incomplete communication to pose as or be taken for successful communication. An awareness of the subtleties of relevance can go far toward helping us avoid these troubles. As a starting point, we can say that communication, as an event, and information, as the content of communication, will be more or less relevant to me depending on its aboutness and the following six factors, all of which are at play in my judgments regarding its meaning :

What I already know (or think I know).
What I think I want and what I ask for.
How I ask for it.
How I apprehend what I receive.
Who I am, what I am doing, and the nature of my situation.
Whether I ask for and get what I desire or what I need.[8]

This list suggests how and why relevance must be regarded as interpretive in nature, and it suggests a different way of thinking about "information" as a theoretical object that may help us to understand and cope with its ambiguities. The question is then recast: Might the apparent contradictions between the physical and cognitive metaphors be resolved by conceiving of "information" as a phenomenon created by the meeting of a potentially informative object and an interpreting subject?

For example, we can assume that a communication whose relevance is in question is a real event whose existence is independent of our perception of it. It may be a text, a speech, or a natural phenomenon, but in any case we can grant that communication possesses a tangible, objective reality of some kind. That this tangibility exists, however, does not ensure that communication will be successful or complete. A third party may establish by independent observation that I speak to you, but that does not mean that you are aware of either the observation or the speech. In turn, my communication to you is relevant to you only if you hear it, acknowledge it, and in some way accept it. It must satisfy a need, reduce your uncertainty about existence, or address and resolve an anomaly. More than anything else, it must have meaning for you, even if that meaning is only to dismiss it (and me). Even so, this dismissal requires that you interpret and make some judgment about my message in the context of your current state of knowledge.

Another idea suggested by this list is that it may be possible to observe the conditions it specifies and so arrive at some empirical and objective basis for understanding how relevance judgments are made. From the perspective of the cognitive metaphor, for example, we can argue that communication is effective, successful, and complete when and if the information transmitted from a sender results in changes in a receiver. It may even be possible to measure the extent and kind of changes that occur, as well as determine extent and way in which information is relevant. The exact unit of measure that might be employed, however, remains elusive. Most work undertaken to understand relevance, by information scientists through research and by librarians through practice, is based on two questions: What factors or elements enter into the notion of relevance? and What relation does this notion specify?[9] It is clear that the elements at play include things and people, technology and ideas, facts and values—all of which manifest the categorical duality and ambiguity that characterizes information as a theoretical object. These features are further revealed in the views of relevance that have evolved over time, but we would do well to remember that unlike biological evolution, in the domain of relevance, few if any early species of ideas disappear entirely.

One way out of the duality is to split relevance in two, assigning one half to the system and the other half to the destination. The system half retains the word *relevance*, but uses it to refer to formal and topical properties that allow a text to be associated with a question. In this respect, relevance becomes a matter of questions, their analysis, the search strategies used to query a collection of texts for the purpose of retrieval, and the relation between system half of the problem and questions. To cope with the destination half of the problem, relevance is replaced with *pertinence* which refers to the properties of texts and users that allow a text to be associated with the satisfaction of a user's need for information.

In other words, relevance specifies and assigns documents to queries, while pertinence specifies and assigns the knowledge in the documents to needs. On the one hand, this distinction allows subjective judgmental factors some room to play by recognizing that a document relevant to a query may not be pertinent to a need. On the other, it makes retrieval effectiveness difficult to evaluate. Relevance is a public phenomenon that can be observed and objectively assessed. A disinterested party, for example, can analyze a document and a question and determine with some degree of accuracy and certainty whether one is about the other, regardless of whether the document in question actually satisfies an information seeker's need. Pertinence, however, is a private phenomenon, and consequently its assessment is necessarily personal and subjective.[10]

While the distinctions between relevance and pertinence, and private and public knowledge respectively are helpful in clarifying the issues at hand, we still face some serious challenges. For most ordinary users of information and information retrieval systems, these distinctions may not be especially valuable or sensible. For them, information is relevant if it does something they desire it to do. Thus, a document that may be relevant but not pertinent will be, for them, merely irrelevant. It would seem, then, that the final judgment of relevance that users make will have a great deal more to do with their sense of what they need after it has been affected by the information retrieved than with anything they may have said about their needs prior to a search.[11]

But what of the information systems and library services which purport to provide people with relevant information? If they cannot do so, they will be judged ineffective by the people whose good opinion really matters, regardless of what else they may do right. In order to determine if our efforts are truly effective, we still must find some way to understand relevance and relevance judgments. Pertinence is a useful concept, but we are still left contemplating, What does it mean? On what grounds and for what reasons do people decide whether or not information is relevant to their needs or not?

A number of approaches to this question are available. We expect our information retrieval systems to help us solve problems that rely on but extend beyond the need to retrieve information. These problems may stem from public policy, business, social relations, or even of the heart; so one very practical way of assessing whether or not information is relevant to us is simply to ask if it contributes to the solution of our problems.

For an information retrieval system is to be judged effective, the information it provides must be useful as well as relevant. Relevance, then, is not only a function of topicality but also of such characteristics as its novelty, credibility, and importance to its user.[12] Still, we might be left wondering if the same information always possesses the same degree of usefulness regardless of the characteristics of particular users and their situations.

While many people, for example, may share the same problem or at least the same kind of problem, it does not mean that the same information will be useful to each of them in exactly the same way. As a result, relevance must be regarded as individual and situational, depending on the user's perceptions, concerns, preferences, current state of knowledge, and view of his or her situation..[13] Of course, this condition also raises the possibility that people may consider an event is relevant to them even, when by any reasonable criteria, it isn't. At this point, from the perspective of a user of information, the conceptual distinction between relevance and pertinence breaks down. Information is either useful or it isn't. That an observer positioned outside of a user's situation might judge information to be relevant but not pertinent to that user doesn't really matter as the judgment is not the observer's to make.

This condition has not caused information science to abandon the concept of relevance despite the fact that more than a few scholars believe that it deserves no more respect than Fairthorne granted the concept of information. Instead, an approach derived largely from cognitive metaphor has arisen on the premise that topicality is a necessary but inadequate criterion for relevance judgments. True, there are many possible non-topical, user-centered relevance criteria that users may employ and that affect their judgment of the relevance of information. In addition, information-seeking behavior appears to be fluid and iterative rather than strictly logical and linear. But this behavior has an improvisational character; and it is at least possible, if not likely, that a user's relevance criteria will change as a search for information unfolds. In fact, the information a user retrieves during a search may cause that user to reevaluate his or her need and abandon one project in favor of another. In order to penetrate and understand the complexity of the reality and private nature of how relevance is constructed in real information seeking situations, we now turn to an examination of research that reveals individuals in actual information-seeking situations. This research is firmly based on the cognitive metaphor and its focus is on how and why information affects its users in the ways it does. It addresses relevance as a cognitive construction rather than as an attribute of informative objects.[14]

Conditional Relevance

We are at a difficult point now in our task of understanding the nature of relevance. Both the physical and cognitive metaphors of "information" helped to bring us here, but we are now entering a territory where their usefulness as maps for the ground we must cover is less certain. The case is clearer for the physical metaphor than for the cognitive metaphor, but for both information is regarded as a determinate phenomenon; and both support solutions that involve assessing cause and effect.

By now we have seen that the system view of relevance, supported by the physical metaphor, simply cannot address the kinds of issues that are central to understanding the complexity and ambiguity of judging relevance. The cognitive metaphor fares better, for if information is a phenomenon that can and does cause a change in the knowledge structure of its user, we can logically conclude that only information that is relevant to a user's anomalous state of knowledge can cause this change. But while this conclusion is comforting, there are three points we must consider.

First, in order to study relevance from a user-centered point of view, we must penetrate the realm of private, subjective knowledge. In other words, we now need to find a way to observe how information seekers actually construct meaning-in-context. According to the cognitive metaphor, we can do so by building models of the cognitive states of information seekers and those of authors as represented in their texts, which we then match so as to retrieve the needed information. But we have yet to figure out how to enter this problem space at the moment that information is used, relevance is established, and meaning-in-context is constructed. The difficulty we face is one of making private knowledge public.

Such an approach to relevance clearly allows for the possibility that meaning, effect, and the relation of information to human needs may each possess an affective aspect. Simply put, relevance may be a product of desire as much as thought, and while the cognitive metaphor implies a place for emotion as an aspect of relevance, it does not explicitly address this issue. Also, the cognitive metaphor posits a universe where individual information seekers seemingly have no intimate connection to one another except through the formal exchange of information. The cybernetic vision of social existence this implies has some appeal as an explanatory metaphor, but it can also be argued that a community is a self-creating entity whose identity is something other than merely the sum of its parts. It may be that in addition to being a psychological construction, rather than a natural phenomenon, relevance is socially constructed. If so, then the once apparently straightforward claim that information is power takes on new ideological meanings.

Information science, whether guided by the physical or cognitive metaphor, seeks to understand, manipulate, and predict the behavior of information. The cognitive metaphor suggests that these goals cannot be accomplished without some understanding of the relations between information and human cognition. If we view information science as a social science, the ambivalence and ambiguity that comes with being human guides information science into areas into which the physical metaphor's notion of science simply cannot go and where the cognitive metaphor must confront implications with which it is none too comfortable. On a more positive note, there is no reason to believe that these two metaphors exhaust the ways in which we can think about information as a theoretical object. It may be that information science is a practice capable of sustaining multiple metaphors.[15] Indeed, as we shall see, the implications of some of the recent work on the nature of relevance may be pointing toward just such a development.

A user-centered approach to relevance once again represents the essential dialectic inherent in information as a theoretical object. If information itself can be regarded as a creation—that of a meeting between a potentially informative object and an interpreting subject—then the relevance of that information to its user must be a similar kind of creation. Just as the physical and tangible reality of a text must be perceived and processed by an act of cognition in order for that text to acquire meaning, the content of that text and its aboutness must be assessed by an act of judgment in order for that text to acquire relevance. Materiality can be accepted as a given, but meaning can only be constructed by and through use. Likewise, the topic of a text may be self-evident, but its relevance to the user can only be constructed by and through its use. Finally, the use of a text and its content will be conditioned by a user's need and the context within which that need arises. More important to the use of information than the need that motivated its being sought is the need that it actually addresses, and in some cases, helps to create.

Once I decide that information might help to address a problem posed by a life situation, it becomes possible to begin constructing an information problem or to use a term introduced earlier—a problem space. This problem space may take the form of an initial anomalous state of knowledge, perhaps represented by a search query, but as soon as the first bit of relevant information is returned in response, that initial state is gone, changed by the effects of the information. The information problem, in other words, changes during and as a result of the search process. The context I bring to the situation and the need arising from that context are changed at each new exposure to relevant information, and each exposure may lead to new criteria of relevance because of these changes. Relevance emerges during and by means of the process of searching for information.

My information need can never really be completely expressed prior to my search for information, nor can my judgment of its relevance be exclusively based on the aboutness of a text; both depend context and situation. Users of information systems are not merely searching for information, for texts, or for aboutness. They are seeking relevance itself. They desire cognitive change.[16] Relevance is what connects a text and its user in a way meaningful to the situation that drives a user to seek information in the first place. We can also see the effectiveness of an information system by dismissing the idea that it exists merely to retrieve wanted texts, seeing it instead in terms of the desired change. What is finally solved may not be what initially motivated the search, but a search for information is nothing less than a search for meaning.

So, if relevance is to be found in the relation between a text and its user as constructed by that user, what is at play in this relation? How is relevance constructed? The six factors identified earlier in this chapter provide a good starting point. Given what we have discussed so far, we must now ask what difference does the use of information make to me? Am I any different after its use? Note that I need not be any better off for using information for it to be relevant. In the presentation of relevance, the only issue is whether or not the use of information will change me, my situation, or both. Presumably the nature of this change would be a desirable one, which is why I was seeking the information in the first place. Information science, however, really does not address the ways that information might change me in undesirable ways (unless it is assumed that being rational, I will always reject as irrelevant any retrieved information that might cause me to change in those ways). But there are reasons to believe this assumption is untenable.

Research in information science reveals a great deal about how and why people make the relevance judgments they do. It is encouraging to learn that a finite range of relevance criteria is normally at play, at least some of which can be associated with the objective properties of texts-as-things, as the physical metaphor might suggest. In studies regarding user responses to retrieved citations and documents, characteristics of the citation and the content of the documents have been discovered to be related to relevance judgments. Nevertheless, what seems to be most important about any of these characteristics is how they are interpreted. As might be expected, citation elements including title, style, the author, the journal title, and, if available, an abstract are all related to relevance. Sometimes a single element will dominate this judgment, or, on occasion, a first impression might be changed.[17]

A title, for example, might present keywords that lead me to identify a document as being topically relevant which I later dismiss based on the abstract. Or perhaps I am unfamiliar with the author or the journal. Or I might be not be sure of the title, but the author and journal convince me to take a look. Perhaps if it had not been quite as recent as it is, I would have rejected it as irrelevant. At any rate, when I do finally take a look at the article, other criteria regarding its content come into play. Of particular importance are the depth and scope of the article and my judgments regarding its validity, clarity, and the appropriateness of its methodology. It might well be that I find the article generally acceptable but not entirely to the point. Therefore, even though these kinds of characteristics are inherent in the text itself, it is clear that I must bring something to the text from outside of it in order to arrive at these interpretations. Other important criteria related to the text as physical entity include its obtainability and cost. A text I can't get except at a cost I am unwilling to pay is as good as irrelevant.[18]

While the text and its objective characteristics are related to relevance judgments, other significant criteria are cognitive in origin, representing what users bring to the text based on their uniquely personal situations. These criteria reveal how the meaning of the text is constructed from their encounter with the text. The application of these interpretation criteria creates not only relevance but, by this very action, information. Remember that within the context of the cognitive metaphor, any form of communication is information only if it actually informs. The communicated file must effect a change in the receiving file, i.e., it must transcend a kind of threshold. It must be noticed and then regarded as important. In other words, it must be relevant to the receiver. In a sense, this means that all information is relevant, and communication that is not relevant is not information. Communication *becomes* information through a cognitive act of interpretation which itself begins with a judgment of relevance.

Let's assume I have a text in front of me. The first thing that must occur in order for me to make information of it or, in other words, for it to inform me, is that I have to be able to place it in the context of what I already know and what I think I need to know. The most obvious failure at this point occurs when the text is in a language I cannot read. While its content may be exactly what I need, the text remains irrelevant; but there can be many reasons why I cannot understand the text. If I can overcome this obstacle, then I can make a preliminary judgment that it is relevant to me. However, as I begin to read the text and attempt to absorb and use its content, I may find that it doesn't make sense. I may think it incorrect, off the point, unpersuasive, or unrelated to my need. In any case, it makes no difference to me. It does not cause any change in the way I think about my need, my problem, my situation, or my world. If, on the other hand, the text leads me to change anything about my thinking, i.e., it makes a difference to me, then the text becomes relevant information.

The effect of change caused by my reading of the text may be a little or a lot. It may even be the case that although I do not find the text to be useful, my exposure to it has changed the way I conceive my need. If asked later whether the text had any effect I might not remember, let alone attribute relevance. After all, I did not use it, at least not as its author intended; yet the text may have set me off on an entirely new way of thinking about my need that allowed me to find other texts that did become relevant information. It might also be the case that I reject a text because I have no context in which to place it, but later, after reading other texts that affect my thinking, find it has become relevant after all. As I mentioned at the beginning of this chapter, we are now in a very grey area of vital importance to both information science and professional practice and one that still eludes our full understanding.

What I already know and what I think I need are part of the context that I bring to a text and apply as criteria to judge its relevance. While the text itself, the objective characteristics of its bibliographic features, its content, and its topicality all provide a starting point for this process, a judgment of relevance also depends a great deal on non-topical, situational criteria that I bring to the text and use to interpret it. These criteria form the context in which a text must fit and which must be changed by the text for it to become relevant information. As they are diverse, we cannot assume they can be applied equally to a given text in a given situation. In addition, their relevance to my interpretation of texts can and do change, depending on the texts and my situation.

I am very likely, for example, to apply criteria that derive from my previous knowledge, background, and experience to judge a text's novelty, importance, and potential usefulness. These criteria contribute to an objective assessment of a text's content, but unless I am a truly disinterested observer, I will also apply interpretive criteria that derive from my attitudes, beliefs, preferences, and biases. In addition, my emotional response to the text come into play. For example, I will likely determine content which is topically related to my need, yet based on a political position or methodological approach I find distasteful, to be irrelevant. I might judge one text relevant because I know and trust its author, and reject another because I've never heard of him or know him and believe him to be wrong most of the time. I might be utterly bored by a text or truly excited by it. In the former case I might reject a text that is nevertheless topically related to my need, and in the latter I might judge a text to be relevant even though it is not quite on target.

Then there is the reputation, visibility, or quality of the source itself. It might be that I find one text more relevant than another because the former's source is well regarded by other practitioners of my discipline while the latter's is unknown to most of them. Note that the application of these criteria may have little relation to the actual content of the text. Finally, there may aspects of my immediate situation that predispose me to favor or reject a text. For example, I may not have much time in which to solve a problem, and a text, while topically relevant, may simply be too big for me to use.[19]

Such criteria comprise the filter I use to determine if a text can find a place in my context, but if a text is to become information it must not only fit into my context, it must change it. We can see what must happen if we reorganize these criteria just a bit. There are three contexts that affect the reception and use of a text. The first, and in the end, most important context, is the psychological, internal context that I bring to the text which reflects my cognitive state of knowledge—anomalous or otherwise. It is constituted by my experience with the texts, the subject, and the problem which are all related to my need. It includes my expertise, my beliefs and values, my paradigmatic preferences, and my research experience. As noted above, I must be able to place a text within this context if any change in my thinking is to occur, and ultimately it is change within this context that will be the measure of the relevance of any text to which I am exposed.

The chances of this change happening depend on other two contexts and the change that must occur within them. The act of seeking information provides an external context, including both the purpose and the stage of my research. Based on this context, I will make judgments about how well the search is progressing and determine whether or not my results are meeting my expectations. This context has much to do with whether I accept a text, reject it, or keep it in a kind of preliminary relevance holding area. These decisions in turn depend not so much on my ultimate goal as they do on the objective of my specific search. Am I looking for comprehensive recall of as many relevant texts as I can find even at the risk of retrieving irrelevant ones, or do I wish a precise pinpointing of texts on a specific topic even at the risk of missing some that might be equally relevant? Another important aspect of this context is the intertextual information environment in which a search occurs. As texts are retrieved, the meaning of one changes and takes on new meaning based on its relation to other texts, both relevant and irrelevant. An outcome of this condition is the possibility that the retrieval of the next text will change the intertextual environment to an extent that once dismissed texts will become relevant and once-relevant texts will be dismissed. The changes I expect to experience in this context include a growing awareness of the information environment in which my search is unfolding, and from that a sharper understanding of the possibilities to inform that it possesses and a movement from a condition of being without information to a condition of being sufficiently informed. No final changes in the first context can occur until this condition is satisfied.

Finally, there is the specific problem context itself and the situation defined by that problem. I normally begin a search for information not with the intention of changing everything I know, but rather with the intention of addressing a specific problem and making sense of my immediate situation as it is conditioned and defined by that problem. This context is structured around the issue of why I am seeking information in the first place. As I encounter texts in my search I consider each in terms of whether or not it will contribute to the solution of my problem. One of the trickier aspects of this context derives from the way these encounters may change the nature of my problem and, in turn, change the criteria by which I am judging the relevance of texts

Given a reasonably stable problem, however, I might decide that some texts are on topic and relevant simply because they seem to contribute to solving my problem. Conversely, I might discover some texts that while not on topic nevertheless strike me as relevant because they suggest new ideas or ways of thinking about my problem. Then again, I might encounter texts that are on topic but irrelevant because they seem repetitive, they add nothing new to my knowledge, or I disagree with their assumptions, approaches, or methods, such that ultimately they make no discernable difference to my problem. Of course, I am likely to encounter texts that are not on topic and irrelevant, at least for the moment. But unless a text can effect a change in my immediate problem context, it is unlikely to effect a change in the internal context that I am using over the long haul to make sense of my world.[20]

Relevance as an Indeterminate Concept

Relevance, like aboutness, is a phenomenon that is recognized when it is experienced, but remains elusive prior to experience. Still, it is possible to make some very general statements about relevance, of which perhaps the most obvious is that relevance and aboutness are so intimately related that treating them as separate phenomena appears dubious even for purposes of analysis. If an event is relevant to my situation, related to the matter at hand, and of concern to me, then it is pertinent to me. The event is about my situation, and its relevance is a matter of the relation between event and situation. Relevance is a kind of sign that connects two or more phenomena that share some kind of similarity of content and meaning. And if aboutness is a phenomenon whose appearance depends on a subjective interpretation of the objective characteristics of information, then clearly relevance must be regarded as a similar kind of phenomenon. It does not merely reside in an event or text. It must be actively created by a judgment about an event or text. This condition, however, contains a problem.

If, by some reasonable criterion, one phenomenon is related to another, but I do not judge it so, then their relevance to one another is not established, even though it exists. On the other hand, if by some reasonable criterion, one phenomenon clearly has no relation to another but I do judge it so, then their relevance to one another is established even where none exists. Finally, consider this dilemma: What if my situation is ambiguous, perhaps even anomalous to an extent that I am compelled to suspend judgment altogether regarding the two phenomena? Where the phenomena in question are an event and my situation, how does this condition affect the relevance of that event to my situation? Is event's relevance necessarily determined in the context of my situation and my perception of both my situation and the event, or is it really a matter of some objective characteristic of the event, regardless of what I think about it?

It seems as if the only thing we can definitively say about relevance as a phenomenon is that it has something to do with the relationship between information and the need for information. It appears or fails to appear in a conflation of relations, ultimately between event and situation, but mediated by such other phenomena, as content, judgment, and meaning. Figure 9.1 illustrates these phenomena at play. However, it should not be taken as a model or representation of a determinate condition manifesting measurable variables. The dashed lines are meant to imply just how ambiguous and vague the relations between relevance and event, situation content, judgment and meaning can be.

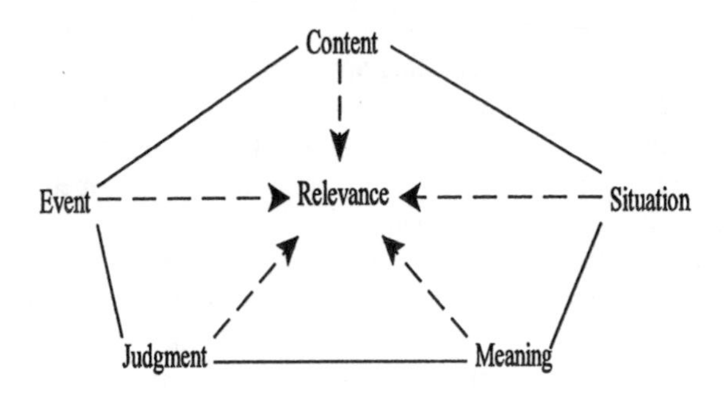

Figure 9.1 The Ambiguity of Relevance

The complexity associated with relevance is also revealed when we consider that each phenomenon at a corner of the pentagon in Figure 9.1 shares a mutually influential relation with each of the other four. The meaning of an event, for example, depends on the judgment made about its content and the situation in which it occurs. That judgment, however, is not free of influences from preexisting determinations of the content and meaning of a situation prior to an event's occurrence. Although we could play this game indefinitely, it is probably not worth it at this juncture. Given the condition of indeterminacy that surround the phenomenon and idea of relevance, we can see why information science has had so much difficulty defining an operational concept of relevance.

The irony of this condition becomes apparent when we consider that this essentially indeterminate concept is central to every task undertaken within library and information science, not to mention every goal. It is arguably the central phenomenon of interest to the information sciences. Every effort to describe, represent, organize, and access texts is an effort to select a signifier *relevant* to what is signified. Every effort to retrieve texts is an effort to find and provide texts that are *relevant* to users' queries at least, if not their needs. Every means of evaluating the quality, effectiveness, and efficiency of an information retrieval system is grounded in an assessment of its ability to retrieve *relevant* and exclude *irrelevant* documents. Despite the apparent straightforwardness of these concepts, especially when viewed from the perspective of the physical metaphor, neither is really fully understood or entirely well defined.

The system view of relevance, supported by the physical metaphor and grounded in its need for unambiguous categories of reality and its measurement, tends to cope with relevance by reducing reality to its essential aspects. Despite producing useful results for practice as well as theory, however, it is clear that this approach is incomplete and fails to address a wide range of information problems related to the outcome of the retrieval and use of information. Conversely, if we are not careful, the user-centered view of relevance, supported by the cognitive metaphor and grounded on its notion that human reality is a psychological construct, can deteriorate into a totalizing relativism by implying that relevance is entirely in the eye of its beholder. Of course if this is true of relevance, then we might as well fold up the tent and go home, as it implies that human communication is more a matter of chance than intention. We must hope that the truth is to be found somewhere between these extremes, if only because neither is sufficient on its own.

Relevance: A Tentative Judgment

At this point we can conclude that relevance—or perhaps more accurately, relevant information—is the product of a particular kind of psychological construction process. It is an outcome of my interpretation of a text—be it a citation, a document, or some kind non-verbal communication manifest by the senses of sight, hearing, smell, or touch. That interpretation will depend on a context that I create from my cognitive and affective store of experience and feeling, my immediate situation and the problem at hand, and the possibilities inherent in the information environment with which the interpretation takes place. We are definitely a long way from the physical metaphor's concept of information, and although much of what has been discussed in this chapter is firmly grounded in the cognitive metaphor, there are enough troubling aspects of this conclusion that we would do well to seek yet another metaphor.

Relevance is a keyword in more ways than one.[21] It represents an essentially contested concept, in that its essence can never be finally identified; and it is and always will be subject to contests over its meaning. This situation is as true for information science which seeks an operational definition for relevance as it is for individuals who attempt to decide if a particular text meets their needs. Relevance can plausibly be determined in a variety of ways, it is contextually dependent, and it evolves as a natural category of an information user's unique and particular experience.

Although shared context is precisely what makes communication and society possible, no two people share exactly the same life context and so no two people will share exactly the same notion of relevance even when confronted with apparently identical situations. I say, "apparently," because my last sentence presumes that all else being equal, no two people can ever share exactly the same situation, the same perceptions, the same thoughts, or the same feelings. So, if they do come to agree on the relevance of a text they both read, it can only result from observable communication and negotiation, and even then, given the essentially private nature of their thoughts and feelings, we can never be completely certain about this agreement. Relevance defies a single, final definition.[22]

This conclusion is troubling if for no other reason than information science still hopes that it will someday evolve from a "soft" to a "hard" science. These hopes, however, are compromised by our inability to offer final and commonly agreed upon definitions of theoretical objects like "information" and "relevance;" their nature as essentially contested concepts, and the elusiveness of any method of measuring the degree and kind of cognitive change that meaningfully represents the condition of becoming informed. On the other hand, the continued existence of cultures and communities suggests that some definite and finite criteria of relevance and meaning are consciously applied by those who seek and use information. That a consensus about the meaning of relevance is unlikely does not render the concept useless. It remains central to the work of information scientists, indexers, reference librarians, and users of information. What the lack of consensus does mean is that the problems information science wants to solve are more complex than we first thought and that the nature of information and its relevance demands that we embrace a genuine epistemological pluralism. There is usually more than one way to get from here to there, and while some paths may take longer, they may also be more interesting. We must remain open to a variety of ways of discovering and knowing.

It is doubtful, given converging advances in computer science and technology that the physical metaphor has yet exhausted its possibilities for discovery. Similarly, the cognitive metaphor, grounded in psychology and the behavioral sciences, shows us that we can accurately measure and predict at least some human behavior. The two combined promise some very interesting work regarding artificial intelligence and its application to both practical and theoretical problems of information. Certainly we are reminded of the fact that these problems are ones of technology and people, but unless we are careful we may neglect a consideration of the ways in which the problems of technology and people are social problems. Understanding information as a theoretical object involves aspects of information as a thing and its interpretation in the context of need, and neither of these phenomena occurs in a cultural or historical vacuum. Jesse Shera's original argument for a social epistemology is based on just this observation.

Common needs and language, manifest in a shared culture, contribute to making information storage and retrieval conceivable, let alone possible, and the individual psychological construction of relevance necessarily must occur within a historical context created by these same phenomena. Informing, becoming informed, and knowing are all matters of individual and social reality. To paraphrase a well known aphorism, it is true that I make my own relevance, but I do not do so exactly as I please. I encounter texts whose material reality and differences from me I cannot deny, but why are they what they are? Do I need these texts because they genuinely satisfy my needs, or are my needs conditioned and prepared by the existence of these texts? Do they create the need they satisfy? My interpretation of these texts will depend on a context created from my cognitive and affective store of experience and my situation, but how does that context itself depend on the material and social context within which I live? The reality of the problem of information and the problem of information need appears to be one of things and people, of individuals and societies, of internal and external realities, and of the moment and of history. It seems little wonder that we know so little about it. At the heart of any metaphor are certain ontological and epistemological assumptions about the nature of reality and how it can be known. The next chapter will pursue this idea further by exploring the notion of information as a social phenomenon.

Conclusion

The work of information science on the nature of relevance alerts us to the essential unity of opposites embodied in the tension between the physical and cognitive metaphors of information. Current work on information need and information-seeking tends to confirm this alert. Both information and the information environment possess a reality—independent of the information user—that conditions how information is perceived and used. On the other hand, it is a meaningless reality until contextualized by a user's cognitive apprehension of it. In effect, information becomes real when it is used, and the nature of its use is inevitably contingent. Thus, the reality and relevance of information, despite its objective form and tangible qualities, depend on what its users bring to it and how they interpret it. But this condition is not the only source of contingency.

Interpretation is an act of negotiation. My need for information may be grounded in my experience, but it emerges from a negotiation I must conduct with reality. I cannot will reality to conform to my wishes. And while my articulation of my information needs may be an outcome of an anomalous state of knowledge, the anomaly itself originates in a pre-cognitive source located in my affective responses to and my relations with the world. These needs then begin as something felt rather than something thought. As of now we really don't know how or why we become conscious of and capable of articulating needs as complex as the need for information. Certainly the social roles we play in the world, at work, and at home are likely sources of information need, but knowledge of the actual moment at which this need takes shape as conscious thought eludes us. It is increasingly apparent, however, that the need for information does not spring fully formed from the mind as did Athena from the head of Zeus. It is a self-conscious construction, dependent not only on what I think about world around me and my place in it, but also on how I feel about it. We need to understand that people seek information not merely for self-directed cognitive ends, but from living and working in social situations with demands of their own.

Given a human reality that is necessarily constructed from the not always knowable or predictable relations between self and others, we must grant that the final goal of information seekers may be as affective as cognitive. Fact may inform me and address anomalies in my state of knowledge, but advice and opinion from a trusted source may ease my troubled mind or put my soul at rest. We might even posit that scientists need affective affirmation as much as a literature review to support their next experiment, just as ordinary people require new knowledge to cope with life's unexpected qualities as much as the visceral information thrill provided by the latest blockbuster action film. To be meaningful, information science must be inclusive. It must focus its attention on a wide variety of information, information users, and information use if it is to assert a legitimate claim to be a science about *all* information and its users. Regardless of context, for any user of information, an essential dialectic is at work. Information is used to make sense of its user's context, and a user's context is what makes sense of information. One can change the other, but it is the their interaction that allows us to make sense of our world.

Endnotes

1. Stephen P. Harter, "Psychological Relevance and Information Science," *Journal of the American society for Information Science* 43 (October 1992): 603.
2. Tefko Saracevic, "Relevance: A Review of and a Framework for the Thinking on the Notion in Information Science," *Journal of the American Society for Information Science* 26 (November/December 1975): 328-329.
3. Joseph W. Janes, "Other People's Judgments: A Comparison of Users' and Others' Judgments of Document Relevance, Topicality and Utility," *Journal of the American Society for Information Science* 45 (April 1994): 161.
4. Carol L. Barry, "User-Defined Relevance Criteria: An Exploratory Study," *Journal of the American Society for Information Science* 45 (April 1994): 150.
5. Taemin Kim Park, "Toward a Theory of User-Based Relevance: A Call for a New Paradigm of Inquiry," *Journal of the American Society for Information Science* 45 (April 1994): 136.
6. Barry, "User-Defined Relevance," 149.
7. Saracevic, "Relevance," 325-326.
8. Ibid., 325.
9. Ibid., 324.
10. Ibid., 331-332.
11. Thomas J. Froehlich, "Relevance Reconsidered: Towards an Agenda for the 21st Century: Introduction to Special Topic Issue on Relevance Research," *Journal of the American Society for Information Science* 45 (April 1994): 128-129.
12. W. S. Cooper, "A Definition of Relevance for Information Retrieval," *Information Storage and Retrieval* 7 (1971): 19-37.
13. P. Wilson, "Situational Relevance," *Information Storage and Retrieval* 9 (1973): 457-471.
14. Froehlich, "Relevance Reconsidered," 125-126.
15. David Ellis, "The Physical and Cognitive Paradigms in Information Retrieval Research," *Journal of Documentation* 48 (March 1992): 49.
16. Harter, "Psychological Relevance," 607, 610-611; and Park, "Toward a Theory," 136.
17. Taemin Kim Park, "The Nature of Relevance in Information Retrieval: An Empirical Study," *Library Quarterly* 63 (July 1993): 333.
18. Barry, "User-Defined Relevance," 153-154.
19. Ibid., 153-157.
20. Park, "The Nature of Relevance," 333-336.
21. Raymond Williams, *Keywords, A Vocabulary of Culture and Society* rev. ed. (New York: Oxford University Press, 1983): 21-25, 252-256.
22. Froehlich, "Relevance Reconsidered," 128.

10

Information as a Social Phenomenon

Need and Use

Conventionally and historically, information science conceives the need for information as a matter of people seeking documents—in some cases documents known in advance, but more commonly documents about subjects related to some need. The purpose of these searches is variously associated with reducing uncertainty, resolving anomalous states of knowledge, or solving problems. As we have seen, the practical goal of information science is to contribute to the design of information systems—conceived broadly as libraries, databases, and the Internet—that are capable of retrieving documents relevant to the queries submitted to them if not to the need that motivates these queries. Ideally, the retrieved documents provide information that satisfies a need. We are, perhaps, close to this goal. We know to design systems of representation based on the anticipated needs of their users. However, we are still learning how to define the relevance of that information with respect to its user's situation.

Unfortunately, we have yet to breach the ambiguity inherent in the representation of documents and in judgments regarding their relevance. Nor can we easily separate the document that is actually desired from the information or knowledge it contains. The fact that the words *information* and *knowledge*, used in the last sentence more or less interchangeably, do not necessarily represent the same thing introduces considerable ambiguity about what people are really looking for when they go searching for documents. I have often used the word *text* rather than *document* to signify the container and conveyor of information because, when people are seeking information, it is often true that they are not searching for documents at all. The use of words like *knowledge* and *document* to signify the goal of a search for information, and the use of words like *retrieval* to describe the acts of seeking and finding information suggest that most of the time people are looking for facts. However, it is more often the case, and often is, that people are looking for advice or opinion. The *text* then, can manifest a variety of forms and purposes.[1]

In their search for facts, people may not use formal information systems at all. Many studies suggest, for example, that libraries are well down on the list of possible sources people think about consulting when they want to look for information.[2] Preferred sources include agencies whose primary purpose is to do things other than provide information, such as banks or retail stores, or other people, such as friends and family. As our study of relevance has shown, values, bias, emotions, and various motives are often in play when texts are perceived and assessed, and there is no reason to suppose that information, including "factual" information, is free of these characteristics. People might seek out information in order to reinforce a prejudice as much as to learn a truth, and information may be produced and distributed to achieve just such an effect.

Sometimes people who seek information fail to satisfy their need. They may use a formal information system when it is not appropriate, or they may fail to use one when it is. Or our information systems may fail them because texts were not represented in meaningful ways. People get facts when they want advice or opinion when they want facts. Sometimes they get facts, but not the ones they want. As we discussed in the last chapter, sometimes they get what they think is relevant to them when it's not really or find something relevant but dismiss it, or they fail because the information provider fails to apprehend an information seeker's situational context and its effect on the seeker's interpretation and judgment regarding the meaning of retrieved texts.

Each of these failures, however, may stem from a common root. It is possible that they result from a failure to adequately understand the essence of information need, information use, and information seeking. As we have seen, the word *information* is not easy to define. As one might expect, the phrase *information need* also raises questions. Why do people *need* information? What does it do for them? Much of the research conducted about the use and users of information is devoted to examining the kinds of demands people make of information systems. In many the cases, the next step is to infer needs from those demands, but this problem is more complicated than it may at first appear.

For example, to understand the use of information, we must first grant that the use of a journal literature by scientists represents only one manifestation of the problem of information. The design and creation of databases of scientific and technical literature, while clearly important, cannot be the only goal of information science. Searching for and using information involves many different kinds of information, looked for in different ways by different kinds of people from different sources and through different channels to satisfy all kinds of different purposes. Any study of information users and use must be specific about who in particular is the subject of the study and what information in particular is being looked for and used. Simply stated, information use cannot be studied meaningfully outside of its specific context.[3]

By now, this idea should come as no surprise. However, the need that motivates these behaviors remains elusive. If information is a thing, for example, what is it about this thing that people need? If its consumption as a thing is compared to other things we consume, such as food, exactly what physical aspect of human existence is sustained by this consumption? If, on the other hand, we view information as a cognitive phenomenon ,whose use accomplishes the reduction of uncertainty or the resolution of an anomalous state of knowledge, does all information reduce uncertainty, resolve ASKs, or solve problems? Is it even intended to do so?

It seems likely that a need for information can be regarded as equivalent to a need for relevant information, and that such a need is necessarily a product of the same kind of relations from which relevance is constructed. It also seems likely that few people seek information as an end in itself, that the need for information is not the same kind of need as other "basic" human needs. Human needs are conventionally divided into three categories: physiological needs, such as the need for food, water, and shelter; affective needs, such as the need for recognition, attainment, power, and love; and cognitive needs, such as the need to make a plan, solve a problem, or understand a situation.[4]

While information may be related directly to the satisfaction of cognitive needs, its link to physiological and affective needs is less than clear. True, these needs are closely related, and they are likely associated with one another in a ways that are difficult to separate. For example, a need for information is not, in most cases, representative of an independent category of needs. Instead, information usually serves as a means to the end of satisfying a more basic need. Such basic needs, on their own, do not automatically motivate a need or a search for information. People who experience such needs may or may not realize that information helps to satisfy them. About the only thing we can definitively say about information needs and information seeking is that these phenomena emerge from the diversity and variety of personal experience and the conditions of human existence.[5] Since human existence is a social existence, "it must also be recognized that these needs arise out of the roles an individual fills in social life."[6]

There is something appealing to the idea that information science might someday be able to identify types of anomalies so that the type of information and information retrieval system best suited to resolving an anomaly can be predicted, developed, and identified despite incompletely defined needs. At the moment, however, the most striking characteristic of the need for information is its indeterminate nature. It is easily confused with other kinds of needs as well as with wants, demands, and desires. Actual measure of its satisfaction remains elusive, and the social nature of its source brings economic, political, and cultural factors into the fray. Information needs may not be, and probably are not, unconditioned expressions of freewill. Rather they are at least partially determined within the context of a dialectic relation between a person and his or her social environment. Simply stated, information need cannot be studied meaningfully outside of its specific individual and social context.

Another View of "Information"

Throughout our examination of information as a theoretical object, a troubling subtext has been present though unexamined. In chapter one, Saracevic's location of the problem of information in an information ecology could have served as the beginning of a very different book about the social nature of information. Shera's assertion that a central goal of librarianship and information science is the improvement of the human condition raises certain moral questions that cannot escape their historical context. References to the intimacy of information and communication throughout suggest that the conception of information as a theoretical object is somehow one of the relations between self and other.

Nevertheless, information science, regardless of the metaphor used, tends to posit information problems and an information reality that are individualized, which in turn limits the extent and nature of its investigations. The language of information science, the words it uses, and especially the way it uses them presupposes a reality in which information systems, information users, and information needs have no apparent history. While the physical metaphor generally assumes a social context for information problems, it does not explicitly address that context. Conversely, while the cognitive metaphor does introduce a social context, particularly through its contextualization of relevance, this introduction occurs only through an appearance of the social in the cognitive processes of individuals. These practices and the use of language that supports them have helped to solve a wide range of practical and research problems, but they also significantly exclude a wide range of phenomena and problems that seem equally compelling. They unnecessarily limit information science by excluding social problems associated with information—limits which may have serious political implications.

The examination of information needs suggests that something greater than individual interest may be at stake. By matching the source of information needs with the roles individuals play in social life, we come very close to explicitly identifying information as a social phenomenon and potentially as a political one as well. By suggesting that information needs are not unconditioned expressions of free will, but at least partially determined by the social/historical identities with which we personally engage, we come close to locating the political context of information. Information science typically identifies barriers to the accessibility of information with system failures. Such barriers include problems of indexing, problems of aboutness and relevance judgments, and problems of retrieval algorithms. While librarianship has a healthy appreciation for problems related to explicit forms of censorship and divisions between information haves and have-nots, it does not always display a sophisticated understanding of the more subtle forms that barriers to access can take.[7] A central issue, though not investigated by many people in great depth concerns the role history plays in conditioning access to information, information seeking, information choice, and the choice of information available. In short, some of the barriers we experience may be due to what our ideas about access and use of information, as expressed by our use of language, allow us to think. In other words, we may be limited by our inability to conceive of alternative possible realities.

Signifiers such as "information access," "information retrieval," "information transfer," and "knowledge management" appear in the information science literature as descriptions of objectively knowable processes; however, their clinical nature and apparent objectivity disguise the way and the extent to which these phrases represent social acts.[8] By constructing metaphors of "information" which focus more or less exclusively on the problems of individuals, we risk not seeing and not understanding how the processes we believe to be related to the problems we wish to solve are socially organized and constructed. Many information scientists insist that their discipline ought to serve practical ends and social needs and that communication and information have implications embedded in social practices.[9] For example, Jesse Shera called for a social epistemology that would embrace the phenomena and regularities of communication practices in society, yet left unanswered large questions about which practical ends and whose needs should be approached first. Similarly, while the claim is often advanced that information science will have a role to play in the creation of an information society, whose information and whose needs will command attention first? Regardless of whether if information science is in a moral or political position to answer the question of whose needs come first, and why, is a compelling one. After all, information is created, collected, and packaged in the full knowledge that it is information and that it is likely to have certain intended effects.[10] What then is the relationship between the social construction of this information and the needs it is intended to satisfy? By privileging the study of certain kinds of information and information uses, and by using a language to justify as well as describe a particular rather than a truly general problem of information, information science may be contributing to that problem as much as it contributes to its solution.

These kinds of questions suggest that if "information" as a theoretical object is confined to signifying only a phenomenon that is related exclusively to individuals and individual use, information science runs a risk of failing to engage the full reality of information. Recognizing this risk, the discipline borrows from other disciplines to investigate the use and behavior of information in organizational contexts. Known as social or organizational informatics, this work concentrates on formal information use environments.[11] Its premise is that the environment in which people work and experience information needs affect the flow of messages into, within, and out of that environment. In return environmental factors influence, if not determine the criteria applied to judgments regarding the value and relevance of information.

Within a formal organization, for example, information need and relevance will be significantly conditioned and determined by both the organization's goals and the structures and methods it has developed to achieve them. While the same general assertion can be made about any form of human organization, work in information science focuses mainly on formal and especially business and government organizations, the rationale being that in industrially developed societies, most of our information needs are associated with work. There are also certain methodological and pragmatic advantages to focusing on work organizations over other types. Their discrete boundaries and specified structures render them relatively open to observation compared to other less formal organizations, and solving their information problems has immediate practical and economic rewards.

Information services and use in such environments are filtered through a context of organizational purpose, organizational culture, and organizational needs. The kind of information needed, the timing of its delivery, and the people who need to receive it are all conditioned by this context; in turn, this conditioning reveals that individual information problems under these circumstances are also social problems. Thus, an organization's communicative style, its ways of doing things, and the affective aspects of its everyday life may be as important to information need and use as are an organization's goals, objectives and strategies.

The study of information in organizational contexts raises questions about the nature and value of different information sources, channels of communication, and content. Who should get what information, when, and why? What is its expected use? How does it actually flow, and how is it actually used? What is the relationship between organizational role differentiation, based on factors such as specialization and expertise, and the development of different classes of information use and users? How do we assure that information appropriate to a given user-in-context is delivered? It is apparent that all of these questions might be asked of any kind of organization including whole societies. Even in the limited context of formal organizations, the answers to these questions are likely to have implications regarding political power and social control. Beyond this context, issues regarding the role of information in societies quickly reveals the political implications of information as a theoretical object. Sometimes, it may itself be a political object.

Arguably, the most intriguing questions concern assessment. Research guided by the cognitive metaphor drew focus to how information might change its user. Work conducted in the areas of relevance and sense-making found the context of information use to be of vital importance in determining an acceptable answer but focused primarily on the way individuals used and were changed by the use of information—an issue of user empowerment. Information was found to be a means by which individuals secure and extend their control over their personal reality. If information is conceived as a social phenomenon, however, we can broaden our scope to discover how organizations, communities, and societies use information to secure and extend their control and power over social reality. What kind of information is used or can be used to extend power in what ways? Whose interests are served? Does relevance possess a social and therefore political character? From the point of view of information social science, the power to set criteria of relevance regarding information needs and use may prove vital to the setting of political agendas and imposition of political solutions. Final judgments regarding the usefulness of information may be based on assessments grounded in values and value conflicts. We may be able to objectively assess the relation between information, its use, and the achievement of goals. But are those goals are worth achieving?

For any social problem to which information is applied, whether as a product or service, there will be problems to resolve concerning how success should be assessed. First, most social problems involve a number of stakeholders. The similarities and differences among their agendas, goals, and definitions of the problem to be solved will influence choices regarding the information needed, the information finally used, and the means of delivering that information. Second, social problems are usually complex, and information is likely to be only one of many resources necessary for their solution. Identifying the role of information in a policy-making process is a difficult task. Any solution to a social problem will require information, but how do we know that information was a critical factor? How do we know what information was important? The role of information is often disguised or hidden in some other process that contributes to the achievement of policy outcomes. Stakeholders' affective response to information may be as influential and decisive as their cognitive and rational responses. In any policy-making process there will be inputs, outputs, and supportive and inhibiting factors in which information plays a role. Third, real world social problems are rarely solved completely. The conditions under which the role of information is investigated are far from experimental, and causality is be difficult to establish.[12]

In addition to these problems, the nature of "information" is itself subject to political interpretation. There is a tendency for information-as-thing, particularly data, to dominate policy discussions. Often this information is strategically deployed to establish the authority of a predetermined position or desired end rather than actually used to illuminate the social problem under consideration. To understand the role of information in social contexts, we must look beyond formal channels and sources. Informal channels of information and their role in social and political contexts are important, diffuse, and difficult to assess, yet they may be of great importance to the final outcome of social and organizational decisions.

Information, and what is construed as informative, can vary with the purposes of its users as well as with the timing and conditions of its use. Information that seemingly should be useful may arrive at the wrong time or in the hands of the wrong person to be of use. Finally, the mere quantity of information available may not be a good indicator of the richness of the information environment. Too much information may be as much of a problem as too little, and the right information at the right place and time is of greater importance than a wealth of inapplicable information. The identity of the *needed* information, however, may vary from stakeholder to stakeholder such that the final judgment may be based on power rather than insight or, possibly, the result of compromise among stakeholders. In a social context then, the relevance of information may be a political category and determined according to political criteria.[13]

Each of these conditions contributes to the difficulties of investigating the role of information in social contexts and especially those associated with assessing the value of information in the process of solving social and organizational problems. At this point in time, information science is just beginning to address such concerns, and some generally useful ideas about how to study them have emerged. For example, it is probably wise to focus on the problems that organizations want to solve rather than what we see as their apparent needs. Information needs are likely to evolve and change as an organization pursues its goals; ultimately the question is not whether an organization retrieved the information it needed, but whether or not it solved its problem.

It is also important to note that the kind of information involved in these problems is not necessarily embodied in texts. An organization may need to create information either by means of original research about its environment or the problem it needs to solve. In some cases these efforts might lead to publication and the creation of a public record, but this outcome varies depending on the nature of the organization and its research. In other instances, an organization may need to capture information from within itself. Most organizations generate observable information about their actions, yet no one may be watching. Tacit knowledge can be tapped to solve problems, but formal methods must be developed to make this information usable.

None of the above should be terribly surprising. It resonates with the idea that the use of information is not an end in itself, and it suggests that sense-making may be as necessary to groups of people as it is to individuals. An organization must make sense of its environment and its relation to that environment just as an individual must. A critical difference, however, is that organizations must also consider their constituents both inside and outside of the organization. Categories concerning both the benefits of information and information use must be established and assessed in relation to the problem to be solved and the effect of the solution on organizational stakeholders: in a word, accountability. In order to grasp the final impact of information and its role in solving social problems, we must discern the effect it has on everyone with a vested interest in the solution.[14] Indeed, solutions to problems may in fact change an organization in important ways, and not always in ways that are anticipated or desired.

Critical Problems

Information science has taken some progressive steps towards understanding information as a social phenomenon concerns regarding this approach remain. Concepts borrowed from other disciplines are accompanied by a sense that if social organizations are not natural phenomena at least they can be studied as if they were. As a result, they become "organizations-as-things" in a way reminiscent of the physical metaphor's view of information-as-thing, and all we need do to contribute to their effectiveness as rational human goal-achieving mechanisms is to analyze them as information-use environments, and then apply what is revealed to their work. The term *information-use environment* itself suggests a neutral category of social existence manifesting identifiable and invariant characteristics.

Still, while there are good reasons for treating such phenomena as natural objects, such an approach fails to acknowledge the social construction of information-use environments. Information and its use sustain the power relations that create and recreate these constructions on an everyday basis, even as they resist other constructions and the power that sustains them. Information may, in fact, be central to struggles both within and between social organizations, to define reality in ways that privilege or challenge the ideas, knowledge, and values that make sense in a given social context. Still missing, however, is an appreciation of information itself as a social and cultural phenomena.

It is apparently easy, for example, to look at an organization and identify certain classes of participants in terms of what they do to achieve the organization's goals. Logic would dictate that different classes of participants have needs for different classes of information, and that in a rational organization, this condition would be known and acted upon. However, as we well know, not all classes of participants functionally necessary to an organization's purpose are equal. In some organizations some classes are "more equal than others"; in others there is no pretense of equality at all. The concept of class can also connote differences in culture and power that reflect organizational hegemonies whose actions in turn result in different levels of inclusion and exclusion among classes of participants. Some participants may be allowed more control over the fate of the collective enterprise than others. Decisions by these participants about who gets what kind of information, when, and how may be the means by which this allowance is created and sustained. Crucial questions arise. Who is in control of the information-use environment? How do they define what is informative? How do they use information, not only to achieve collective goals, but to maintain organizational identity and control?

Words like *information*, *information needs,* and *knowledge*, do not necessarily represent neutral, unproblematic referents but perform ideological labor within organizations and societies.[15] At first, it may appear as if the cognitive metaphor can effect a rescue of the situation; if an individual's cognitive structure or mental image of the world can be studied, perhaps a social organization can be approached in the same way. Bernd Frohmann, however, offers a critique that raises some very troubling questions about the limits of the cognitive metaphor regarding its application to information as a social phenomenon.[16]

Frohmann explicitly addresses issues of power and control associated with what he identifies as the cognitive viewpoint, which we have called the cognitive metaphor. Frohmann claims that the cognitive viewpoint assimilates human communication and information problems into processes involving natural objects and natural events. For him, this effort is a rhetorical strategy that allows the cognitive viewpoint to assert neutral definitions of objective characteristics of reality whose scientific study yields objective knowledge. In the process, however, certain phenomena of potential interest to information science are ruled to be outside of the domain of knowledge. In other words, certain aspects of information and its use, particularly as social phenomena, are not allowed to be investigated.

According to Frohmann, the cognitive viewpoint asserts three essential knowledge claims. It integrates autonomous aspects of information science and information retrieval into a single theoretical framework; it reduces human motivation for information production and use to a single cognitive/cybernetic model of human behavior; and it posits a reality characterized by a radical individualism. Through these claims the social world is reduced to an individual and internal image of the social world. Society and social organizations, then, are nothing more than individuals writ large, and the social and historical forces that may account for apparently stable theoretical objects are marginalized. Through the use of rhetorical language that only appears neutral and objective, a wide range of information phenomena, including anomalous states of knowledge and information-use environments, are deprived of both their history and the political and economic contests that give rise to them.[17]

The knowledge claims asserted by the cognitive viewpoint act like axioms and serve much the same purpose as the exemplar Cranfield experiments do for what David Ellis called the physical paradigm. They allow the problem of information to be reduced to a manageable level by limiting the number of variables that we have to think about, and they point to relations between these variables that are also limited in both number and kind. They allow the beginnings of a science of information but in the process may also limit us to study only certain aspects of the problem by privileging those aspects while excluding others from consideration.

The knowledge claims advanced by the cognitive viewpoint that make its work possible can also be read as ideological claims in a society where historically information, information needs, and those institutions that manage the production and distribution of information are socially constructed and contested on behalf of specific interests. The counter-claim is that the apparent neutrality of information science masks how information as a social phenomenon is embedded in politicized social and cultural practices. If information and communication are regarded as exclusively individual and psychologically internal phenomena, then shared public practices become ineligible for study, thus erasing the social from theories of information science is complete, suggesting that information needs operate according to natural and discoverable laws. What goes unexamined is whether these needs might be culturally and politically created and imposed in order to advance the interests of some participants in the social order at the expense of other participants. By excluding questions of this nature, information science fails to recognize that information needs may be the outcome of an uneasy process by which we accommodate and resist socially constructed identities. Nor can it cope with the idea that disparities between information haves and have-nots arise from the social construction of internal psychological realities in which individuals experience and know themselves as subjects of dominance.

Some of the work on relevance, especially the work of Dervin and Kuhlthau, points in this direction; however, here too the individual is the unit of analysis. Instead, Frohmann, echoing Shera, asks us to take up society as the unit of analysis. From a social perspective, the word *knowledge* is problematic. It is often deployed to form terms such as *knowledge structure*, *knowledge management*, and *knowledge economy*; terms which not only signify material realities but can also be used as rhetorical devices. Under these circumstances, such terms lose their power as theoretical objects. Frohmann argues that information science also uses such terms rhetorically, not to illuminate reality but rather to secure the legitimacy of the "natural" theory of information science .[18] Consequently, this rhetorical strategy restricts the discourse of information science to an instrumental reason that privileges efficiency, standardization, and predictability as the dominant values and goals of information and communication systems. Questions about what ends are to be achieved by these means are either not asked or their answers are assumed to be a matter of widespread consensus.

Frohmann concludes that the cognitive viewpoint in information science actually represents a discursive economy that allows some currencies to circulate while suppressing others. Specifically, it commodifies a particular relation between knowledge and power by describing information, information needs, and information use as natural objects of a natural process of production and consumption. In effect, individual information behaviors are seen as the result of the working of an invisible hand of information rationality as opposed to, for example, political conflict. Thus, the cognitive viewpoint, rather than contributing to a complete understanding of information as a social phenomenon, restricts our understanding by displacing its social aspects and constructing the information user's identity as analogous to capitalism's consumer of goods. The conventional wisdom of capitalism stresses that the market offers goods and services in response to an aggregation of individual demands such that balancing supply against demand will result in an efficient and rational allocation of social values. By extension, the role of information science would be to develop and contribute knowledge that supports the "market" of information, resulting in the efficient management and control of information production and consumption.

It is possible that information science will contribute to the creation of an information society but not in the way that many information scientists intend or believe. Even if information science as an organized academic discipline maintains a strict neutrality toward issues of information politics, the kind of theorizing in which it indulges in may be used by others to design telecommunications and information technologies whose use will contribute to the social construction of human beings as consumers of information. If such networks create needs for commodities for which they are the only source, then questions regarding the social and cultural implications of this condition must be asked.

If we can assume that culture, values, and human desires are embedded in such technologies, then the social and the technological cannot be regarded as autonomous realms of human existence. Technology and culture mutually condition one another's development. Thus, a culture will tend to produce the technologies it needs to become what it believes it wants to be, and those technologies in turn will contain possibilities for ordering social action that encourage some actions while inhibiting others. In a very real and material way, our social and personal identities are reflected in and shaped by the technologies we deploy.

Now, let's assume that we live in an information society, characterized by a confluence of technology and culture that is reflected in the development of telecommunications and information technologies based on the physical metaphor's ability to control and manipulate information as an object and informed by the cognitive metaphor's ability to model and configure human thought and behavior. Seeing as our social relations, and particularly our economic relations are increasingly mediated and configured by these technologies, the information they simultaneously provide to us and draw from us tends to configure and determine the routines of our daily life and our ability to negotiate those routines.

Frohmann, among other observers, suggests that a possible outcome of these conditions is a postmodern way of life.[19] The autonomous ego, conventionally known as the self-determining person, is displaced and mediated by information that flows in, seemingly unrestrained, from a bewildering number of directions. The resultant identity is no more than a terminal of multiple networks and servers. Under these circumstances, identity becomes abstract, variable, and decontextualized, prepared to change and accept the next information-commodity offered as a means of defining and confirming identity. In short, we become the information we consume; because this commodity is of such an ephemeral and inconstant nature, we are compelled to become someone new from one moment to another. At the same time, we are so conditioned to the nature and pace of this kind of change that we scarcely notice it or even experience it as such

There are two ways we can consider this condition. One way is to view information technology as liberating us from unwanted hassles and distracting, irrelevant choices. This technology might make possible the deployment of selective dissemination mechanisms capable of delivering information we need even before we know we need it. According to this view, we are evolving a society in which pluralism, difference, and diversity are realized and celebrated by micro-markets for goods and services. The consumer is in charge. Producers and providers of goods and services are even now constantly telling themselves and us that they must use information and information technology to work smarter, that they must customize their offerings and their benefits to meet a constantly changing consumer demand, and that they must do this to remain competitive. And in the long run, it is competition that yields more consumer choice and, thus, makes us all better off. In short, information science and technology will contribute to the development of a new and liberated mode of consumption.

The other view assumes that information science and technology can be used to constitute subjects through interpellations, or the calling of names. When we are called by certain names, for example, we respond. When called by names that we do not recognize as "me," we do not. An information economy presupposes an information-rich environment, which allows for the creation and use of many different names by any given individual, even if for that person these names are not entirely congruent and, in some cases, are actually contradictory. Nevertheless, by the consumer choices we make, and the relevance feedback we provide through those choices, we willingly give providers of commodities the names by which we can be called. Compared to other kinds of commodities, our use—or at least our consumption of information—is especially revealing in this regard. By keeping track of what names we respond to when called, we leave a record of an identity which can stored and used by someone else. An image of each of us as a consumer of particular information products can be constructed—our identities reduced to artifact status , merely a peculiar search algorithm, that is created and recreated with each search we conduct for information. Subjectivity becomes constituted and manifest by dialog with a machine that stores a record of that dialog and so a record of our subjectivity. Our identities as consumers can be established, and these identities can be traded. They can be bought and sold on an identity market, if you will, and shared among those seeking both our identities and our business. All in all this condition promises to introduce greater predictability and stability to consumption and the market.

In an information society, then, the ultimate commodity may not be information, but identity. The record of my consumption of commodities, especially information, becomes the index of my self, and this record can be rendered accessible to others, including corporations and governments. From this surveillance will come offerings—customized and prepared just for "me." Questions still remain. Are goods and services customized for me as an individual, or am I customized as a consumer of goods and services? Is my identity now still mine and self-determined, or have I relinquished this autonomy to a system that is making choices for me while disguising it as convenience? It is important to realize that this use of information technology would expand consumer choices beyond those available in mass markets. Customized products and services are likely to address real needs and really satisfy them. The issue turns on some troubling questions. On whose terms does this occur—mine or the system's? Is the exchange truly a fair one? What do I forgo by accepting these conveniences? What choices am I not allowed to consider because I never know about them? In a postmodern information society there seems to be little doubt that my life will be easier but at what cost to my freedom? At issue is who controls the naming and the calling—the consumer or the provider of consumer commodities?

The resemblance of this description of postmodern life in the information society to real life is arguable, and there are reasons to believe that even if such forces as identified here are at play, human beings will find ways of resisting them. On the other hand, this dystopian information future is not entirely implausible. Can information science be used to contribute to this dismal future? Is it worth worrying about? The Internet and the search engines designed to retrieve information from it represent exemplary engineering applications of the basic findings of research conducted within the context of the physical metaphor. And what of the technology already in existence that tracks and records my movements as I navigate my way through cyberspace? It is also possible to employ the findings of research conducted within the context of the cognitive metaphor to design technologies that can use the records of my searches for information, their results, and anything else that might indicate my relevance judgments regarding the information I retrieve; the resulting model could then be used to draw inferences and make predictions about my personal goals and desires. Though we seek to maximize choice, it may that instead we are creating the illusion of choice. Some would even argue that instead of celebrating diversity, information technology is being used to disguise a genuine, universal human identity and to privilege the individual at the expense of human community. Either way, the true meaning of freedom is more elusive than ever.

Examples of Critical Problems

Of course, we are not yet in a position to know with certainty whether either optimism or pessimism is warranted, but clearly we are living in a situation in which we need to make individual and collective sense of what is happening to us and what we are doing to ourselves. If nothing else, however, it would appear that we are wise not to treat words like *information* and *knowledge* as neutral categories of empirical analysis. Even though we do not yet have clear understanding of the problem of information as a social problem, it is not difficult to identify current examples of the kind of social problem under discussion. *Transportation*, for example, may appear at first to be a quite neutral term having to do with how we move things and people from here to there, but is it really such a straightforward concept?

Like the need for information, the need for transportation manifests both general and particular aspects. Communities have a general need to move people and goods about in order to sustain economic relations. In this sense, transportation can be regarded as a public good. Everyone benefits from its provision, whether directly or indirectly. This general need, however, is mediated by and manifest in a unique individual need to go to different locations at different times and for different reasons. In this sense, the provision of particular forms and instances of transportation can be regarded as a private good whose use yields private benefits. Thus, there can exist a variety of ways of providing transportation that will satisfy both the general and particular aspects of the need for it; but the means actually chosen are configured and conditioned by a variety of historical and cultural interests.

In the United States a number of conditions can be identified that have influenced the way we provide transportation for ourselves. First, American culture exhibits a strong commitment to the values of individual independence. We also have a history of business and corporate interests that have played a significant role in setting the social agenda regarding transportation issues. To a great extent, the automobile industry owes its prominence to the cultural values of individual independence, and its growth is associated intimately with the growth of the oil industry which has interests of its own regarding transportation.

The development of road building and contracting industries is also a part of this mix, and the success of these industries is in turn related to the way government has chosen to provide for transportation as a public good. Taken as a whole, each of these conditions has contributed to the developmental spread of our cities that requires that we have roads, automobiles, and the gasoline to power them. While this oversimplified account hardly does justice to the history of American transportation, it at least points to some of the reasons why the United States became what many call a car culture. Perhaps another way of saying this that reflects our concerns about information is that we have become an automobile society.

Now let us consider the "road not taken." Arguably, American culture has favored the particular aspects of the need for transportation over its general aspects, and the road not taken representing the reverse of this situation is mass transit. If the general and collective nature of the need for transportation had been granted priority, and the resources and capital we now have invested in the present transportation economy had been devoted to mass transit, we might have evolved a system of transportation that could address individual needs in the context of the general need. We also might have evolved a very different culture. Instead, the general and collective need for transportation is marginalized, and the existence of the automobile creates the need for automobiles. We need a way to get around from home to malls and schools and from our suburban living sites to our urban work sites.

The question that we typically face as we confront this problem is not what kind of transportation do we need, but what kind of car will we buy. In this context, the individual is configured as a consumer of automobiles rather than as a user of transportation. Ultimately, and despite the incredible variety of vehicles made available in the market, choice is reduced. We accommodate to these conditions by purchasing an automobile, and the insurance, gasoline, repairs, and through taxes the roads that accompany the automobile into our lives. Resistance is possible. One can choose not to have a car and to rely on public transportation, but this option comes with certain obvious disadvantages for personal mobility. For some people—those of limited incomes—this choice is not a matter of resistance but a matter of necessity. In effect, the disadvantages of not owning an automobile are imposed upon them, such that it is hardly fair to call it a choice at all.

Let's now turn to information. In American society, the values of independence, accumulation, and property fuel the development of an information technology and an information economy that allows and encourages a freedom to manage personal financial security in a way unimaginable just a few years ago. This condition also allows some to argue that we now have the means available to end expensive government programs such as social security, reduce the tax burden on ordinary Americans, and return to individuals both the responsibility and freedom to do as they choose with their money. The Internet provides a variety of sources of investment guidance, opportunities, and means to participate in a growing information economy. By redirecting funds from low earning, government sponsored savings programs and into the private sector, it then becomes possible to enhance the value and growth of the economy as a whole in a way that everyone will benefit.

Consider, however, the burden this development might impose on ordinary individuals who must, under these conditions, manage their own financial fate. Great effort would be required in order to become sufficiently informed to make wise choices. The time required to do appropriate research and keep track of one's investments is itself a formidable cost. Each of us would have to become sophisticated seekers and managers of information. We would have to effectively collect, store, and organize a great quantity of accurate information about our own financial status and goals, about business and the economy in general, and about particular investment risks and opportunities. To do so would require a high level of information and telecommunications literacy, which itself would be difficult and time consuming to acquire. In effect, an obligation to participate would be imposed, and the consequences of failure would be great.

But what if a modest but guaranteed financial security for each of us was conceived as a collective rather than individual responsibility? Under these circumstances we might regard ourselves as being free from the need to live a life whose central focus is money. We might regard ourselves as free to pursue a variety of goals centered on any number of different values—for example, family or spirituality. While information would be equally important to the pursuit of these goals, it would be a very different kind of information, and different kinds of institutions responsible for providing information would be required. Once again, it is possible to see how terms like *retirement* and *financial security*, like *knowledge* and *information transfer*, do not necessarily represent neutral categories useful merely for the analysis of a social problem. Rather, contests over their meaning represent an ideological aspect of the problem to be analyzed.

Neither of these examples is persuasive on its own, and both represent admittedly quick takes on complex problems with rich histories. Still they point to some very compelling questions about where needs come from, how they arise and are expressed, why some needs are privileged over others, and why some means of satisfying them are favored over others. The role of conflicting political values and interests in the determination of human needs cannot be easily denied. As we develop our information society we should ask: What kind of information, information sources, and information institutions are being prepared for us to use? What kind of information environment already exists, and is this one that we always want to have? Are there choices that we have failed to consider, but should? Is there a role for information science regarding the identification of the choices we have and the analysis of the choices we are making? Clearly, it is not difficult to accept the notion that we need information, and that as the movement toward an information economy accelerates, we will need an increasingly large amount and diversity of information merely to be able to participate in society on a basic level. Why do we get the kind of information we do, and is this the information we really need?

The idea that the categories and concepts that information science conventionally uses might not be objective, neutral, or natural phenomena challenges the discipline to rethink their meaning and the purpose of their application. It may be that these discursive forms act to naturalize phenomena that are not givens of nature, and that instead they play ideological and political roles in positioning human beings as consumers of information already prepared for them. Concepts and methods of research that naturalize theoretical objects in this manner also naturalize the social order implicated in their construction. Social reality and the information practices that sustain a particular reality as a given natural order are simply accepted; questions about the way ideologies, power, political interests, and deliberate distortions contribute to that reality and those practices go unasked. Problems, including those of information, are paradigmatically defined within the constraints of a status quo that does not allow dangerous questions to be asked.[20]

If social relations, including relations of power and domination, constitute and are constituted by communication and information technologies, then naturalistic assumptions that human subjects and technical artifacts can be treated as stable theoretical objects available for disinterested observation have to be challenged and discarded. Instead we have to consider how dominant social interests are embedded in and configure social relations by means of information technology. For instance, the apparent democratization of access to information that can be made within the capitalist context of the production and distribution of information may, in fact, yield some genuine benefits for some individuals. But whether these benefits are an outcome of design or accident, the essential structure and ends of the information environment remain unchallenged. The result more often than not is access to information that someone else wants us to have, and in the process our consent to the system of production and distribution of information as social value is manufactured and validated. Under such circumstance, our freedom may be an illusion.

Are these concerns valid? Perhaps not, but do we want to risk the easy assumption that ours is the best of all possible worlds? Does the role information science plays in the development of an information society include an obligation to deconstruct the information practices by which the information age is created and the identities of information producers, distributors, and users that people it? Terms like *information, knowledge,* and *need* carry both referential and ideological meaning and so represent positions from which meaning can be controlled and from which the direction of social change can be determined. Who is in control, why, and in what direction are we going? These seem to be important questions to ask.

If the problem of information is conceptually constrained as merely an individual problem, then we cannot ask questions about whose individuality takes precedence and who is privileged by our information systems, our information infrastructures, and our information practices. We cannot ask questions about how cultural and historical conditions and differences of interests and power may lead to an assertion that a particular interest is universal, and we cannot inquire as to how the general need for information may be distorted into particular forms that deny individuality even as they appear to celebrate and enshrine it. Unless we grasp and understand information as a social phenomenon, we will not be able to penetrate and understand the information ecology whose actions and interactions determine the meaning of the keywords Saracevic identified as the central concepts of a science of information in chapter one.

Conclusion

This discussion does not mean that we should cease thinking of information as an individual phenomenon, but it does pose an interesting dilemma for information scientists. How can we follow Dervin's lead, reject the ideal of treating individuals normatively, shift our focus to the needs of individuals as they are actually situated in their unique sense-making contexts, address Frohmann's task of rejecting the individual's internal experience as the exclusive source of information needs, and engage the social nature of information? Can we resolve the apparent contradictions between the need to focus on individual sense-making on one hand and collective sense-making on the other? To do so requires that we arrive at an understanding of how information systems have been configured in order to configure individuals as consumers of information as a commodity and of the material constraints of social reality in the configuration of individual cognition. Thus the work of both Dervin and Frohmann provides a starting point for answering these questions by suggesting that many of the central assumptions of information science are actually metaphors posing as objective reality.

Information as an individual phenomenon must be examined in the context of its social nature. We must grasp the idea that information systems are configured on the basis of a social knowledge that includes information science. As Saracevic suggests, information science does play a role in the creation of information societies, but how our knowledge and the systems based on it attempt to configure individual subjectivity in predetermined ways are what privilege some subjects over others. The systematic nature of these predeterminations mean that some (and seemingly always the same) individuals find our information systems congruent with their personal sense-making efforts, some maintain a balancing act of accommodation and resistance, and some find no congruence at all between their need to make sense of the world and our information systems. This dilemma posed here will not be an easy one to resolve. Not all of what information science has accomplished so far is irrelevant to the task, but parts may be standing in the way of engaging let alone solving the dilemma. Of all the compelling questions that information science as a discipline now faces, perhaps the most pertinent is what role the discipline wants to play and ought to play in the creation of an information society.

Endnotes

1. T. D. Wilson, "On User Studies and Information Needs," *Journal of Documentation* 37 (March 1981): 3.
2. Robert S. Taylor, "Question Negotiation and Information Seeking in Libraries," *College and Research Libraries* 29 (May 1968): 181.
3. Wilson, "On User Studies," 4-6.
4. Ibid., 7.
5. Ibid., 7-9.
6. Ibid., 9.

7. Michael H. Harris and Stan A. Hannah, *Into the Future: The Foundations of Library and Information Services in the Post-Industrial Era* (Norwood, N.J.: Ablex Publishing Co., 1993), 85-102.
8. Norman Roberts, "Social Considerations Towards a Definition of Information Science," *Journal of Documentation* 32 (December 1976): 249-250.
9. Gertnot Wersig and Ulrich Neveling, "The Phenomena of Interest to Information Science," *The Information Scientist* 9 (December 1975): 133-134.
10. Roberts, "Social Considerations," 253.
11. Robert S. Taylor, *Value-Added Processes in Information Systems* (Norwood, N.J.: Ablex Publishing Co., 1986): 34-47.
12. Michel J. Menou, "The Impact of Information—I. Toward a Research Agenda for Its Definition and Measurement," *Information Processing & Management* 31 (1995): 455-477; and Michel J. Menou, "The Impact of Information—II. Concepts of Information and Its Value," *Information Processing & Management* 31 (1995): 479-490.
13. Menou, "The Impact of Information—I," 467-472.
14. Ibid., 473-475.
15. Bernd Frohmann, "The Power of Images: A Discourse Analysis of the Cognitive Paradigm," *Journal of Documentation* 48 (December 1992): 368-371.
16. Ibid., 380-384.
17. Ibid., 372-376.
18. Ibid., 376-378.
19. Bernd Frohmann, "Communication Technologies and the Politics of Postmodern Information Science," *Canadian Journal of Library and Information Science* 19 (July 1994): 13-17; and Gernot Wersig, "Information Science: The Study of Postmodern Knowledge Usage," *Information Processing & Management* 29 (1993): 229-239.
20. Frohmann, "Communication Technologies," 13.

11

Semiotics for Information Science

On occasion a scholarly work reaches far beyond its intended audience and alters the course of thought in a number of different disciplines. Ferdinand De Saussure's *Course in General Linguistics* is such a work[1]. The Library of Congress subject headings assigned to this work include "language and languages," "linguistics," and "comparative linguistics." While it is certainly about these subjects, its influence has reached philosophy, literary criticism, and the social sciences. It is a seminal work for the emerging field of cultural studies and, as we shall see, information science.

At first glance it seems as if Saussure's work should be a central pillar of information science, yet with the exception of a few forays here and there, the discipline has not embraced the implications of Saussure's suggestion that a science of signs is possible.[2] Saussure named this science "semiology," and identified it as the study of "the life of signs in society."[3] More recently, this discourse has come to be known as semiotics. Representation, and the relationship between representation and what is represented, are at the heart of both semiotics and information science. Both semiotics and information science are vitally concerned with representation and the production of culture. Both are concerned in different ways with signs used for the purpose of communication. Both address issues of what we know, what we could know, and what we have forgotten.

Regarding our understanding and use of knowledge and other kinds of cultural products, both also ask similar questions. Given the signs with which it is possible to communicate, why is one chosen rather than another? What rules govern the process of choosing, ordering and use of these objects?[4] How do the rules governing the constitution of signs as cultural products affect and condition our possible choices? Why are some relevant to understanding situations and others not? In a given speech situation, why do we deploy some signs but not others? What kind of culture is produced as an outcome of our choices? Do the rules ever change and if so, how? In sum, both semiotics and information science are concerned with relationships between content and its representation, between signifier and signified, between reference and referent, and between informative objects and their meaning.

The problem of retrieving information relevant to a need is essentially a problem of penetrating and understanding the nature of what Saussure called the "sign." Signs appear to us as apparently unrelated and heterogeneous objects, yet they are necessarily linked by a common bond created by an act of reading.[5] As an illustration, Roland Barthes provides the following list: "a garment, an automobile, a dish of cooked food, a gesture, a film, a piece of music, a piece of furniture, a newspaper headline."[6] Each, he says, can have something to say us. It might be something about the social status or lifestyle of the owner of the object. It might be an idea or a political statement. Even the notion of reading a newspaper headline, while seemingly obvious, offers more than one way to read, including reading between the lines. And while our reading of other kinds of signs may occur, even without us ever realizing that we are reading, it is the nature of the sign to be at least potentially informative, available to be read, and open to interpretation.

Some signs are straightforward. We read the red light at an intersection as an instruction to stop. However, the apparently simple sign on the interstate highway informing us that the next rest stop is sixty miles away may trigger a complex set of associations regarding the urgency of arriving at a destination, how long the children in the back seat can hold out before a bathroom break, and the price of sanity on a long distance automobile trip. Other signs, such as a symphony by Mozart, a painting by Picasso, a novel by Pynchon, or an article in the *Journal of the American Society for Information Science and Technology*, are considerably more complex at the outset and demand greater attention if they are to be read at all.

Barthes argues that we routinely accept many of the signs we encounter at face value and that "we take them for 'natural' information."[7] This "taken-for-granted" aspect of signs, however, disguises the way signification provides the principles by and through which meaning is determined and social reality is created and shared. In short, nature may not be quite as natural as it appears, as Barthes himself says.

> To decipher the world's signs always means to struggle with a certain innocence of objects. We all understand our language so "naturally" that it never occurs to us that it is an extremely complicated system, one anything but "natural" in its signs and rules: in the same way, it requires an incessant shock of observation in order to deal not with the content of messages, but with their making.[8]

The sign—or as we might say in information science the informative object—is never as what Barthes calls innocent, if only because its role in a system of signification and meaning that includes a great many other signs. In addition, signifier and signified share a complex relation, as is true of the relation between a text and its content. An article in a scientific journal, for example, may be quite informative on its own, but it is only a part of a body of literature, an ongoing discourse regarding a given subject; and its full meaning depends not only on what it has to say, but its place relative to other statements in the discourse. The rules governing that discourse are likely to be complex and perhaps contested in ways that allow alternative readings of the article in question. Understanding exactly what the article signifies—in other words, what it is about and its relevance to its reader—is neither an easy nor a straightforward task. A surplus of meaning can intrude upon interpretation.[9]

This difficulty affects both ends of the information retrieval process. On the front end is the difficulty of assigning an accurate and adequate representative description to a given text, most commonly but certainly not exclusively a document, in order to appropriately place it in a system of texts whose organization makes a given text accessible when we want it. There are also possible tensions between the signification of a text and the system of organization applied to its description and control, especially if there is no appropriate place in the system to put that text. On the back end is the difficulty of assessing whether or not the accessed text is actually relevant to the need that prompted its retrieval.

Information science and semiotics share another important characteristic: the central theoretical object of each discipline bears an unmistakable indeterminacy. Saussure observed that unlike other sciences, whose theoretical objects are given in advance and then examined from a variety of viewpoints, linguistics presumes "that it is the viewpoint that creates the object."[10] A word, for example, is nothing more than a sound arbitrarily associated with the expression of an idea, such that the same idea can be expressed by a variety of sounds, with no final criterion for determining which word/sound best expresses it. Thus, the study of language engages a number of dualities that necessarily reflect certain unities. For example, there can be no speech without thought, yet without speech, thought will find no articulation; it has both an individual and a social side. Even though individuals use language to speak and so to think, speech has no meaning unless language is itself a social institution. To fix attention on one side of these dualities, Saussure argues, would lose a consideration of the other. Similarly, to focus exclusively on either information's material or cognitive aspects risks overlooking an important part of its reality.

Saussure concludes by saying that speech, and so communication, is a combination of physiological production of sounds, a physical transmission of sounds, and a psychological association of sounds with sound-images that signify concepts or ideas; a combination which must occur in a social context that associates the same sound-images with the same concepts for most speakers.[11] Speaking (*parole*) is individual, willful, and intellectual—a code that allows speakers to express their own ideas; in contrast language (*langue*) is a social phenomenon, with a history and existence independent of any given speaker, while passively assimilated by speakers who share the same culture.[12] Meaning, the associative element, is created in the moment of speaking a language, thus uniting of *parole* and *langue*.

The parallels between language and information are striking. Saussure's work suggests that text can be regarded as something akin to *parole*. It is willfully created by an individual who wishes to communicate with others. It is unique, a product of choice, and almost unlimited with regard to what it might be. The actual number of ways in which words, musical notes, images, or colors can be put together in order to express an idea has not and likely cannot be counted. The content of a text, on the other hand, is much like *langue*. It is a social phenomenon, constrained by history and culture and serving as the shared set of concepts and meanings from which texts are constructed. For example, a text that is composed from "information" that is not shared in some way by writer and reader will be incomprehensible to the reader.

Unfortunately, as information science lacks the equivalent of Saussure's distinction between *parole* and *langue*, the word *information* must do double duty, signifying both speech (regardless of its medium) and thought (both text as wells as content). This condition contributes to theoretical confusion in information science and is at the heart of Fairthorne's frustration with the word. Saussure was aware of an irony regarding his own use of certain words to refer to certain ideas when he wrote "all definitions of words are made in vain; starting from words in defining things is a bad procedure."[13]

Parole and *langue*, like information, have tangible qualities, and they can be reduced to conventional written symbols, but they also represent the collective storehouse of concepts and ideas that provide us with something to talk about.[14] When Saussure argues that writing is merely sound images in tangible form, he comes close to saying that text, regardless of its medium, is merely the tangible form of information. Information then, is the sign that unites text and content.

But what is the nature of this sign, and how can it help us to understand the nature of information as a theoretical object? If definition is inadequate as a means of relating things to words, what procedure is recommended? Saussure's answer begins with the idea that language is not constituted merely by the process by which a word corresponds to the thing it names. Such an approach, for example, would also assume a one-to-one correspondence between the subject of a text and a word that names that subject. If ready-made ideas existed in the world simply waiting to be named, this approach might make sense, but unfortunately, the linking of a name to a thing is not that simple. What, for example, of such subject-naming words as *democracy* or *love*?

Rather, says Saussure, the linguistic sign is a "double entity," uniting not a thing and a name, but a concept and a sound-image.[15] This sound-image can take the physical form of a spoken or written word, but the forms themselves are entirely arbitrary and by themselves without meaning. They merely stand in for a sound-image that psychologically realizes a concept. The sign then is a two-sided psychological entity in and through which concept (which he calls the signified) and sound-image (the signifier) are intimately united. Each recalls the other (Figure 11.1).

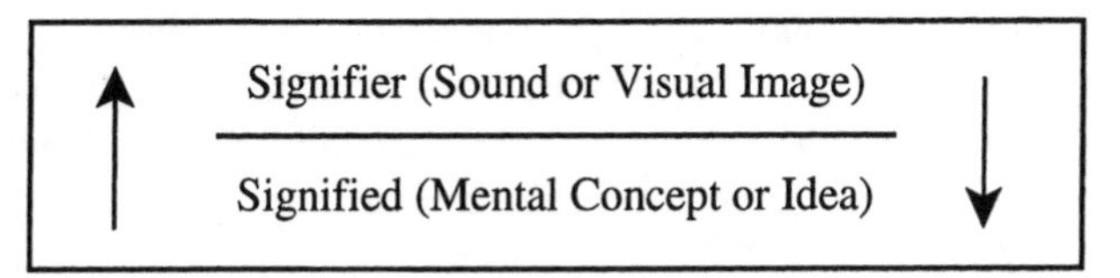

Figure 11.1 The Sign

The signified is an idea or mental entity grounded on some referent in the social or material world. The signifier is the pointer or the signal of the presence of that idea and its deployment in discourse and communication.[16] According to Saussure, the sign displays two "primordial characteristics." First, it is an absolutely arbitrary construction. The signified is not linked to the signifier by any inherent relation between the two. For example, there is no reason why *a tree* should not be named *un arbre*, as indeed it is in French. There is nothing about the idea or the reality of a tree that determines the way in which it is signified. Second, the signifier is linear in nature. In unfolds in time, as sound in speech, and as a spatial line in writing. To receive the communicative message of the signifier, we must wait for it to unfold the signified.[17]

Again, the affinity of the "sign" and the "informative object" as theoretical objects is striking. To receive the communicative message of a text we must wait for it to unfold its content. There is also an arbitrary quality in the relation of text to content—that is to say, many different ways of saying the same thing. Outside of the language of mathematics, there is no absolute determinate relation between text and content. This one exception, however, reveals a disjunction in the above analogy. The relation between text and content can never be entirely arbitrary as is that between signifier and signified; as soon as the relation between the latter two linguistic elements has been fixed in the sign, any given sign in a text will convey a certain relatively a priori fixed content. It is precisely this condition that makes possible the organization and control of texts for the purpose of access.

Still, the informative object, in a manner similar to the sign, realizes its existence by means of a unity created through a relation between text and content. The former acts as the signifier and the latter acts as the signified. Texts as signifiers are themselves composed of signs, and so texts must also unfold their content in time. However, while the representation of content within a text necessarily conforms to Saussure's notion of the arbitrary quality of signifiers, the representation of texts for the purpose of access cannot follow this precept. The assignment of an index term to represent a text and its content, for the purpose of retrieval is a second order signification, and it cannot be arbitrary. On the contrary, the index term, as a signifier, must be selected on the basis of an a priori logical and semantic relation to the text it will signify. Indexing languages, represented by thesauri, are consciously and deliberately created to avoid arbitrariness and ambiguity as much as possible. Without this kind of control, information remains elusive, disorganized, and, as a result, inaccessible. Now we truly confront a dilemma. If we are to consider the sign and the informative object as being the same kind of theoretical phenomenon, we must resolve the apparent contradiction between the former's essential arbitrary nature and the essential need to control the latter.

Saussure's insight regarding the simultaneous mutability and immutability of the sign provides a way out of this dilemma. The sign is an arbitrary creation because any sound-image or word can be used as a signifier, yet it is also fixed. Thus, language appears to a speaker as a given. It is determined by a community of speakers who share a language and sustain it historically by means of convention and tradition. Each speaker passively receives it as a routine and everyday matter of childhood. In other words, signs, are relatively unchanging despite their essentially arbitrary quality. At the same time, the sign "is exposed to alteration because it perpetuates itself," and over time results in "a shift in the relationship between the signified and signifier."[18] Language is a social institution, says Saussure, and its arbitrary quality is precisely what opens it to change, but such change is likely to be slow and more likely to occur to signs whose meaning can be culturally contested.

The informative object, like the sign, is relatively immutable. Although arbitrary in the sense that the signs used to compose a text are essentially arbitrary in nature, once the selections are made and the text composed, it remains fixed and will not recompose itself. While a new edition of a book may be published to replace a prior one, it is a *different* text. Similarly, certain words can be lifted from a document or chosen from some other source to describe the contents of a document because of their generally accepted meaning and their relation to the content of a text.

At the same time, we recognize that information, if not the informative object, can experience change over time. An object that was once informative becomes obsolete and loses its power to inform. New information is created as the conditions of existence change and new referents are created, and informative objects that once meant one thing now come to mean something else. In these instances, the relations between texts and their content manifest change. As a signifier, the text remains constant, but as a signified, the content changes as the viewpoint brought to bear on the informative object changes. This changing relation between text and content and between signifier and signified constitutes a change in the meaning of the informative object, as new meanings are assigned to existing objects. This simultaneous immutable and mutable quality of the informative object allows the possibility of second order representation for the purpose of organization and access. This quality also implies that while indexing languages must necessarily change over time if they are to adapt to the way information changes, this change will occur slowly so as to allow the ordering and control of information, much as a culture orders and controls the meaning of signs.

In other words, information, constituted by informative objects is like language constituted by signs. Both are social institutions subject to the same social forces and same kinds of change that can and do result in changes of meaning that reflect changing realities. Yet each also displays a relative stability that offers the possibility of its control. The meanings manifest in and expressed by both unfold and change over time.

The Role of Time Regarding Meaning and Value

Time has another role to play in signification, and this role and its implications reveal another essential attribute of the sign and its affinity with the informative object. In order to create and express meaning, signs engage value as well as signification. The meaning of a sign as an instrument of communication depends simultaneously on signification and value, on the relations between signifier and signified within the sign, and on relations between signs. Like signification, value and, therefore, meaning unfold through time by means of exchange. According to Saussure, language manifests "a system for equating things of different orders."[19] This condition implies that exchange relationships exist allowing the substitution of one thing for another. As in economics, for example, one commodity can be exchanged for another; in language, a signifier is "exchanged" for a signified. The sign's function of linking *parole* and *langue* through such an exchange makes possible our ability to form sentences that we have never before spoken and to understand sentences that we have never before heard.

This condition implies that language manifests two axes of reality. In Figure 11.2, modified from Saussure, *AB* represents the axis of simultaneities and manifests the relations of co-existing things. (Note that the intervention of time is excluded from this axis of reality.) *CD* represents the axis of successions; while only one thing at a time can be considered, it is the axis on "which are located all the things on the first axis together with their changes."[20] Taken together these axes describe a dual reality constituted first by a system of substitutable values, and then by a system of values that are interrelated with respect to time. Saussure uses this idea to make clear that language has both synchronic and diachronic aspects. In order to understand the given state of a language, one must ignore its diachronic aspects. In order to understand how a language is changing and has changed, one must ignore its synchronic aspects. Both of these tasks cannot be undertaken at the same time.

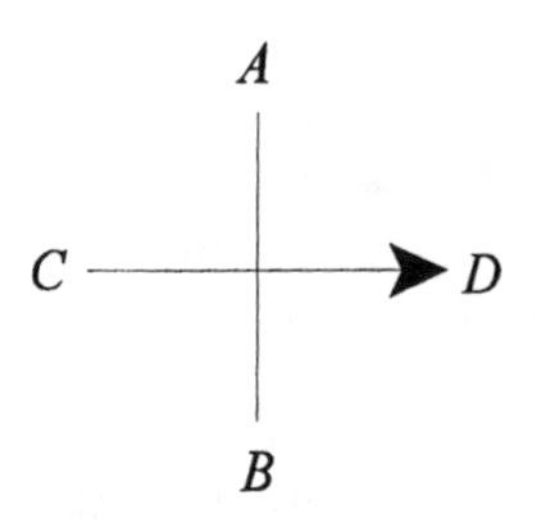

Figure 11.2
Simlutaneities and Successions

This condition of language recalls O'Connor's application of synchrony and diachrony to information. As text, information displays a synchronous character. It manifests itself as information-as-thing, possessing certain invariant attributes whose presence qualify any object as potentially informative and intrinsic to the text itself. It also allows a relatively objective description of the topical aboutness of a text along the lines of Hjorland's notion of true statements that can be made about a text. In turn, the synchronous nature of texts allow them to be classified according to their synchronous attributes, and this condition immutability allows one text to substitute for another in the sense that all texts paradigmatically assigned to the same classification are presumed to be about the same subject. Each text is a specific instance of the general representing category.

As content, however, information demonstrates its diachronous character and its mutability. This is not to say that the content of a text literally changes over time. Recall that once a sign fixes a relation between signifier and signified, by convention, a certain stability of meaning is established. A book using certain signs to speak of animals will not somehow become a book about furniture, but as we are about to see, the value and so the meaning of that content can change as a function of the perspective from which the text is viewed. Certainly time can cause such a change in perspective. A novel one reads in high school may reveal entirely new meanings when reread as an adult.

Time, however, is not the only factor that can affect a reader's perspective. Intertextuality is also at play in two ways. First, just as a given in an ensemble of texts changes from the addition of new texts and their content, the role played by the content of any given text in the ensemble may change between the extremes of seminal and obsolete. Second, as readers come to know the content of more texts about a given subject, their assessment of the value and meaning of the content of any given text they already know may change.

To apprehend information as an object of control for access, we must engage the synchronous aspects of information. We cannot anticipate changes in the value and meaning of content; and even if we could we might easily find ourselves in a state of utter inconsistency and confusion resulting in a complete breakdown of control. In order to account for the nature, behavior, and representation of information at any given time and to design and execute systems of organization and access, we cannot concentrate too closely on how the meaning of information changes. But control of information has another aspect manifest in the judgment of its relevance. For that purpose we must ignore the synchronous aspects of information and instead engage the diachronous and intertextual relations of content in order to understand why the meaning of a text changes.

Roland Barthes insight regarding how the value of a sign is established helps to resolve this dilemma. To begin with, we have to understand that a sign is not an abstraction but a real object. It is a concrete entity whose existence is determined by a material as well as cultural and psychological association between signifier and signified, such that if only one of these two linguistic elements is retained, the entity of the sign vanishes.[21] Attempting to understand the sign by exclusively focusing on but one of its elements runs the risk of mistaking a part for the whole. Saussure writes that "[a] succession of sounds is linguistic only if it supports an idea."[22] The same can be said for the succession of elements that make up a text or a succession of texts, for that matter. Without a signifier, however, the potential signified remains an abstraction of pure thought. Again, Saussure reminds us that concepts "become linguistic entities only when associated with sound-images; in language a concept is a quality of its phonic substance just as a particular slice of sound is a quality of the concept."[23] If the sign exists only through the relation between signifier and signified, then likewise, information exists only through the relation between text and content—or in other words, through the relation between its physical and cognitive aspects.

Both the sign and the informative object consist of two intimately and inextricably linked elements, but this link alone cannot define their reality. Both must be delimited and related to others of their kind before they can be defined and understood. Both unfold their role and meaning over time. A foreign language, for example, does not immediately or explicitly reveal how to analyze its sounds. To do so, the speaker must know the meaning and function of particular sound-images in the context of their relations to other sound-images. Saussure writes,

> just as the game of chess is entirely in the combination of the different chess pieces, language is characterized as a system based entirely on the opposition of its concrete units. Language then has the strange, striking characteristic of not having entities that are perceptible at the outset and yet of not permitting us to doubt that they exist and that their functioning constitutes it. Doubtless we have here a trait that distinguishes language from all other semiological institutions.[24]

However, contrary to his final statement, there may be at least one other semiological institution that manifests the same trait: information.

Consider Saussure's discussion of the identity of the sign by means of the example of the 8:25 PM Geneva-to-Paris train. What constitutes the identity of this "train"? It is always, everyday, the same train, even though it is very unlikely that it is literally composed of the same locomotive, the same coaches, or the same personnel on every trip. The key to the train's identity lies not in its inherent material elements, but in the differences between it and other trains as manifest in a system of "trains" composed of routes, schedules, leaving points, and destinations. Still, one cannot conceive of the train's identity outside of its material realization. It is not merely the idea of a train that travels from Geneva to Paris, but the value of "train" given by what it signifies and by its relative position to other "trains." Saussure explains this condition by reminding us that,

> language is a system of pure values which are determined by nothing except the momentary arrangement of its terms. A value, so long as it is somehow rooted in things and their natural relations . . . can to some extent be traced in time if we remember that it depends at each moment upon a system of coexisting values.[25]

In other words, the meaning of a sign depends not only on the relation between signifier and signified, but also on an identity and value based on its relative position in a system of signs—in other words in a language constituted by other signs.[26] This conclusion suggests that the value and meaning of an informative object, for example a text, depends on both the relation between text and content and on an identity based on its relative position in a system of texts, i.e., in a discourse constituted by other texts.

Information and the Sign

At this point, some tentative conclusions are in order. The affinity and kinship between the informative object and the sign and between information and language as theoretical objects is based on the fact that all informative objects are necessarily signs ultimately expressive of a relationship between a signifier and a signified. The sign, although a material object, is always much more than just that object. It is also a psychological and cultural entity. Saussure makes this point when he writes that linguistics "works in the borderland where the elements of sound and thought combine; *their combination produces a form, not a substance.*"[27] The fox running across my lawn in Tennessee is *le renard* scampering through a field in France, and of course neither word is the animal I see. Both signs are merely forms that represent a substance. Only a community can create a language. The meaning of a sign in the form of either a spoken or written word is entirely a product of convention, common usage, and social acceptance. Still, we must remember that language is also a system. Its elements—its sound-images and concepts—derive from a system of elements whose role in the system of language necessarily depends upon their relations to one another. Meaning, whether of sign or informative object, cannot be understood outside of a context determined by its intertextual relations with other signs.

Like linguistics, information science works in the borderland of two elements: text and content. This borderland is the terrain of aboutness, representation, relevance, and their contribution to the organization of and access to information. In this borderland, an informative object, its content, and its meaning meet as a necessary step in the determination of its relevance to an information user's need. As with the linguistic situation, this situation engages an exchange of values. For access, in particular, content is first exchanged for text, then text for representation. This process represents the signification of aboutness. Access, however, is complete only when need is satisfied. To do so a need must be exchanged for a query signifying that need, which in turn is exchanged for representation to produce retrieval of information.

From the perspective of an information user, a mirror image of this process then occurs. At the moment of retrieval, representation is exchanged for text; upon reading, text is exchanged for content and if access is successful, content is exchanged for relevance and the satisfaction of need. In short, content is exchanged for knowledge. From the point of view of a user of information, aboutness and relevance are themselves merely different moments of the same phenomenon. For example, we predict, on the basis of a match between a query term and an index term, that retrieved informative objects will manifest an expected value we call relevance. However, if a system of information organization and retrieval fails to signify the relevance of potentially informative objects to its user, that system has failed. In the final analysis the determination of aboutness and relevance is both mutual and reciprocal, and the product of this relation is information. Information, to be relevant to its user, must be about the user's need.

To abandon the argument at this point, however, is to overlook the basic fact that information relevant to a user must be meaningful as well, which in turn recalls the notion that meaning is a product of signification and value. However this condition is one that information science implicitly addresses, but tends not to fully grasp. On the one hand, the physical metaphor emphasizes the materiality and immutable characteristics of the informative object as both a theoretical construction and an object of control, which is somewhat like treating the signifier alone as the sign. The cognitive metaphor, on the other hand, embraces the arbitrariness and mutability of the informative object and recognizes that aboutness, and as a result relevance, must be constructed from extra-textual imports, and this is analogous to treating the signified alone as the sign. Both metaphors make the same mistake of taking the part for the whole, although each directs attention to a different part.

Information is more than either of these metaphors convey on their own. It is the product of the relation between them, and like language, it is a system of signs whose values owe their existence and measure "to usage and general acceptance."[28] Its life depends on a community of users, non-users, and producers that agrees a given object is informative and, in turn agrees about its content and meaning. In other words, information science might well be regarded as a form of semiotics. It must embrace the relations of signification and value that produce meaning including relations between text, content, and representation. It must also embrace the relations between texts existing in a system of texts.

Finally, let us consider Saussure's notions of linguistic value, after which we can engage Barthes's insight regarding Saussure to arrive at an understanding regarding the nature of informative value and its relation to meaning. A crucial property of the word, spoken or written, is that it stands for an idea. In other words, a word as a signifier posits a value, which is clearly an element in signification, such that any given sign is itself, potentially, a counterpart of other signs. The value of a sign depends not only on its general use and acceptance, but also on the simultaneous presence of the other signs.[29] Thus, language, speech, and meaning are constructed from relations between the signifiers and signifieds that constitute signs and the relations between signs, as illustrated in Figure 11.3. The relations that constitute *langue* and *parole* as communication, and assign values to each sign deployed in a discourse are the syntagmatic relations that exist between *A*, *B*, and *C*, the paradigmatic relations that exist between *A* and *A**, *B* and *B**, and *C* and *C**, and their interaction with one another. The simultaneity of

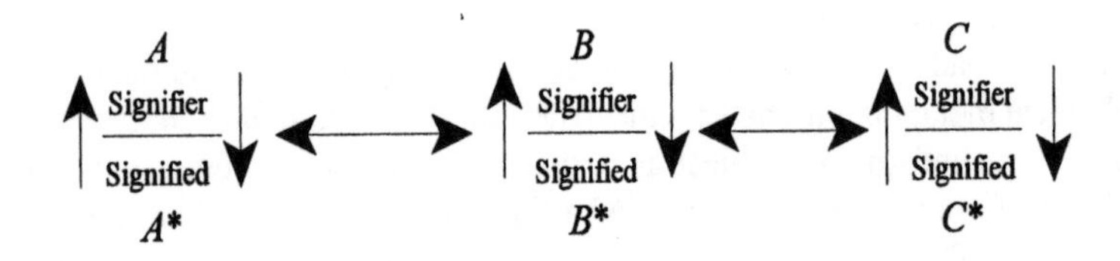

Figure 11.3
Syntagmatic and Paradigmatic Relations

syntagmatic and paradigmatic relations is the source of meaning.

All values, even those outside of language, are always composed of dissimilar things that can be exchanged for one another as well as other comparable things. Saussure writes,

> Both factors are necessary for the existence of a value. To determine what a five-franc piece is worth one must therefore know: (1) that it can be exchanged for a fixed quantity of a different thing, e.g. bread; and (2) that it can be compared with a similar value of the same system (a dollar, etc.). In the same way a word can be exchanged for something dissimilar, an idea; besides, it can be compared with something of the same nature, another word.[30]

From a conceptual viewpoint, value and signification although intimately related, are not at all the same thing. The value of a word, functioning as a signifier within the system of language, depends not only on its signification but on its opposition to other words. In other words, the content of a word and the value of a sign "is really fixed only by the concurrence of everything that exists outside it."[31] The fullness and essence of the sign and the meaning it expresses depend as much on opposition as affinity, on difference as on similarity, and finally on the reciprocally determining relations of within and between signs.

From a material viewpoint, as noted earlier, this conclusion implies that "it is impossible for sound alone to belong to language."[32] The sound-image is a necessary tangible element that supports value and meaning, but on its own it is arbitrary and meaningless. It merely differentiates between one signifier and another. The same condition holds true for letters and their combination in writing. Distinct sounds, sound-images, and words in print constituted by combinations of letters are all entirely arbitrary with respect to signification, serving only to mark the difference between one sign and another.[33]

The above has profound implications for informative objects. While absolutely necessary to the process and communicative purpose of informing, objects are arbitrary, serving only to mark off differences between one meaningful instance of communication and another. This condition is as true of documents as it is paintings, pieces of music, buildings, and any other object that human beings by convention agree upon and, by means of that agreement, constitute as being informative and culturally meaningful. This not to say that particular attributes of such objects do not imply their informative potential and cannot be used to organize and control them for the purpose of access. Their meaning, however, is not due to these attributes in and of themselves, but rather depends on the relations that exist within and between them in a system of signs and signification. As Saussure says,

> in language there are only differences. Whether we take the signified or the signifier, language has neither ideas nor sounds that existed before the linguistic system, but only conceptual and phonic differences that have issued from the system. The idea or phonic substance that a sign contains is of less importance than the other signs that surround it. Proof of this is that the value of a term may be modified without either its meaning or its sound being affected, solely because a neighboring term has been modified.[34]

In sum, signs, including all forms of informative objects, do not stand on their own declaring their meaning and signification as independent entities. The substance of both language and information remains elusive. Everywhere and always, the relations among and between the material objects that constitute each necessarily mutually establish and condition one another's existence and meaning.

The Semantics of Informative Objects

In an essay entitled "The Semantics of Objects," Roland Barthes begins to explore this phenomenon in a way that has profound implications for information as a theoretical object.[35] His project is to extend Saussure's semiology by trying to understand how humanity gives meaning to things that are not exclusively linguistic in nature: How do such objects signify? At the outset, Barthes grants "a very strong sense to the word *signify*:"

> we must not confuse *signify* with *communicate*: *to signify* means that the objects carry not only information, in which case they would communicate, but also constitute structured systems of signs, i.e. essentially systems of differences, of oppositions and of contrasts.[36]

This point raises some intriguing implications regarding such deliberately informative objects as texts and for information science as the study of information in its textual form. It means that texts as objects, whether composed of sound-images (words) or other kinds of images (pictures, music, etc.), not only communicate, but signify. Like any system of signs, they too constitute structured systems of differences—oppositions and contrasts. Their meaning cannot be understood outside of a context created by the system of which they are a part. To study information then requires that we engage the informative object's place and role in a system of informative objects and its meaning in the context of that system. We must also be aware of how that meaning can change as a result of changes in the system, including the addition of new objects and the withdrawal of old ones.

Barthes's description of how we conventionally define objects resonates with the way information science tends to approach and discuss texts as objects of utilitarian value except that his purpose is to critique this approach. Objects appear to us in the world as things ready-made to fulfill a function, their use-value apparently self-evident. These appearances may be misleading.

> Ordinarily, we define the object as "something used for something." Consequently the object is at first glance entirely absorbed in a finality of use, in what is called a function. And, thereby, there is, we feel, a kind of transitivity of the object: the object serves man to act upon the world, to modify the world, to be in the world in an active fashion; the object is a kind of mediator between action and man.[37]

Substitute the word *text* or *document* for *object*, and Barthes will have provided us with a very serviceable definition of *information*. Although his concern in this essay is for the meaning of nontextual human fabricated objects, his argument reveals the nature of the informative object which is of central concern to information science. The function of an object, however, even that of an apparently functionless character, is always at play in its signification and value. In other words, objects are imbued with meaning.

Upon perceiving an object, including a text, we are always faced what it means and what it *means*. The first sense of meaning is a product of how an object signifies its function and is related to its literal and obvious use. It is derived from the attributes that characterize, identify, and are determined by its function and use. These attributes in turn imply an object's signification, the reasons for its use, how it is apt to be used, and the conventionally expected ends that its use obtains. On the other hand, the second sense of meaning is a product of its syntagmatic role in a system of meaningful objects. (At this point, Barthes recalls Saussure's discussion of the meaning of signs, and extending its implications to meaningful objects in general.) The syntagmatic role of an object alerts us to other objects and other meanings to which it is systematically related, and reminds us that when objects are combined in certain ways, they can compose a message, a story, and a meaning that no one of them, or even all of them if viewed serially and without connection, can convey. While each object is unmistakably informative, another meaning and a different kind of informative quality arises from their combination. This intertextuality holds for texts that constitute a discourse as it does for words that constitute a text, and it makes possible the logical organization of texts for the purpose of access.

Bathes uses the following example to illustrate his point.[38] The appearance of a telephone by itself conveys a certain meaning attributed to it by its function. It is a mechanism of telecommunication, although even here one must remember its essentially arbitrary nature as a sign. To someone of a culture that knows nothing of electricity, let alone telecommunications, the function of a telephone and its signification will elude understanding. Even if such an understanding can be taken for granted, however, the context within which the telephone appears must be considered. If it resides on a stand next to a bed and matches the decor of the bedroom, it implies domesticity, a certain kind of taste, and perhaps a desire on the part of its owner to be easily available to friends and family. If it appears on the desk of a teacher or a corporate executive, style is likely to be less relevant, and function more so, although in each instance these functions will differ. Similarly, the absence of a telephone in the bedroom might imply that I don't wish to be disturbed while I'm sleeping, while the absence of one on my desk at work might imply that I'm not important enough to warrant a private means of communication with the outside world. In each case, more information is needed in order to fully determine the content and meaning of the scene. The point is that the second meaning of the object, the one beyond its immediate signification, depends not only on its function but on *something else*: the cultural, and perhaps ideological, role played by the object in a system of objects, such that it serves to signify its user as well as itself as a sign.

In information science we frequently encounter the same kind of phenomenon. Two texts, for example, may be "about" the same subject. They have a similar if not identical function and are intended to accomplish similar if not identical ends. From the perspective of the physical metaphor, they might very well be treated as equivalent and represented in a retrieval system by the same index terms. There may indeed be very many good reasons for this choice, not the least of which is that the same words extracted from the texts occur in the same order of frequency for both of them. In effect, they can be regarded as belonging to the same paradigmatic order and they are substitutable. From the perspective of the cognitive metaphor, what matters is the difference between the texts, especially if their relation is one of opposition regarding the their subject. Two users, based solely on the literal aboutness or function of the two texts, might find both relevant to their need for information. Conversely, it is entirely possible each will arrive at a different conclusion regarding the texts' relevance or irrelevance. One source of this difference will be each text's unique and different interpretation of the subject. Another will be the categorical differences that separate the users themselves. The nature of the use made of each text by each user signifies something about those users and their differences, and about the different role each text plays in the discursive formation of the subject in which our two users are participants. The meaning of each text, despite their similarity of aboutness, is revealed only in the opposition present in their intertextual relation. For both of our hypothetical users one text (and not necessarily the same text) is "right" and the other is "wrong."

Now let us consider two articles on the labor theory of value. It is plausible that each will be indexed in a similar if not identical manner, given that the function of each is to explicate the theory. As informative objects, each can stand as an equivalent of the other, yet one may be an expression of classical economics and the other an expression of Marxist economics. While each is likely to be signified by index terms that point to theories of value and Marxism, their reader, depending on his or her own politics, might view one as evidence of a wrong-headed Marxist ideology and the other a dissimulating bourgeois apology. In other words, the full essence and meaning of each article will stand revealed only when its place and role in a discursive formation regarding the labor theory of value is determined with respect to a reader's interpretative context, despite indexing that is neither inaccurate or inadequate when the articles are considered exclusively in terms of the objective attributes of their aboutness. The meaning of signs, including texts, depends on the relation between signifier and signified within them and their opposition to and affinity with other signs.

This task is less difficult for some signs that it is for others. Recall Saussure's point that the value of a sign depends upon its general social use and acceptance. The practice of science, for example, arguably represents the most rigorous social method of disciplining discourse. The signs and their relations by which scientific discourse is constituted are fixed by theories and operations precisely as a means of fixing their meanings to the greatest extent possible. Signifiers such as "velocity," "distance," and "time" in the discourse of physics are associated with signifieds that are fixed by theories and the operations that measure the realities they represent in way that admits slowly to challenge. Science, however, is not the only human activity whose discourse manifests this stability. "Marriage," for example, is a sign whose value is fixed not only by law but solidly reinforced by conventional behaviors and cultural practices. As we move from definition to experience, however, we begin to lose control over the value of signs. Does "time" mean the same thing to a child as it does to a dying man? For both child and man, a second hand on a watch will click sixty times to yield the passage of a minute, but the experience of that passage may mean one thing to the child and something very different to the man.

In some domains of human activity, convention may be contested or even nonexistent. Widespread use and acceptance of the value of a sign may be difficult to come by. A political scientist might theoretically and operationally define a democracy as a polity in which government officials are selected by means of free and fair elections under conditions of genuine competition. It is equally possible that even under these conditions the same people always win, then use their official positions, to enrich themselves and their supporters. Does the reality of democracy depend at least as much on the ends achieved by a government as it does on the means by which it is constituted? What is actually meant when a speaker describes a nation-state as a democracy? And depending on the actual nature of that state, what might this statement signify about its speaker?

Clearly, convention is ambiguous regarding signification and value, a situation which has profound implications for the representation and relevance of information. Given that the conditions for which convention is an unreliable guide, resulting in the use of a sign whose signification and value is contested, how do we determine what is being spoken and how is being spoken about?

This problem resembles what Barthes calls the "sign-function."[39]

> The sign-function bears witness to a double movement, which must be taken apart. In the first stage (this analysis is purely operative and does not imply real temporality) the function becomes pervaded with meaning. This semantization is inevitable: *as soon as there is society, every usage becomes a sign of itself*; the use of a raincoat is to give protection from the rain, but this use cannot be dissociated from the very signs of an atmospheric situation.[40]

We are again dealing with the issue of what a sign means and what it *means*. The raincoat's function is apparent and gives meaning to it as an object, but it also speaks of how we react to a situation. At the very least, it says that getting wet is not a desirable condition. If we investigate the material and the style of the raincoat, we can learn even more about the person wearing it and the culture within which he or she lives. As a sign, it unfolds its meaning.

Information science can benefit from Barthes's analogy by recognizing that texts are not only composed of signs, they are themselves semiological signs that constitute semiological systems we know as discursive formations. Thus, text as signifier and content as signified are intimately linked to produce a kind of sign called an informative object, whose value and meaning are called information. Such signs are utilitarian and functional, in that they imply a use and a signification, yet they also manifest the double movement of Barthes's sign-functions.

Within the discourse they constitute, their meaning depends on their role in that discourse as well as their self-contained characteristics.

Information then is manifest by two languages speaking simultaneously. A first-order language, operating within the context of convention, expresses the function of a text and what it has to say about a subject in order to communicate, inform, and convey knowledge. However, an easily missed second-order language, speaking about how a subject is spoken about and why is also present. This second-order language is crucial to the value of a text—its meaning as a sign—because it relates to cultural, historical, and ideological meanings that are not necessarily explicit.[41] It is possible that producers of texts do not intend for these meanings to be explicit. It is possible that producers of texts are themselves unaware of the implicit meanings they impart to their texts. And it is possible for a user of a text to read a meaning into it that is independent of its producer's will.

As noted earlier, science seeks to avoid such problems through vocabulary control, although this discipline can and sometimes does break down. During times of paradigmatic contests that Kuhn argues are central to scientific revolutions, it is possible to find scientists using the same *parole* to, in effect, speak different *langues*, or as Barthes would put it, the same "lexic" is deciphered differently. Two early seventeenth century astronomers, for example, might observe the "movement" of the stars across the night sky. For each the signifier is constant, yet one takes it literally to mean that the stars are moving while the other employs it metaphorically to describe an apparent action that results from the fact that the earth is moving. When each observes the same phenomena and even agree on the objective nature of their observations, they differ regarding the meaning and theoretical implications of their observations. In the social sciences, not to mention other discourses that make even greater use of metaphorical signs, the possibility of an incongruence between the first-order and second-order language that constitutes a text, between its *parole* and its *langue*, is exacerbated. As a result, such texts and the discourse they in turn constitute are difficult to control for the purpose of access, as their aboutness and relevance to a reader are shrouded by an inherent ambiguity. An informative object then is likely to be a kind of sign for which the impossibility of dissociating and differentiating between signifier and signified, as much as it is to be desired, will not be easily realized, and information itself is likely to remain an indeterminate theoretical object.

In its approach to "information" as a theoretical object, information science tends to deploy two dominant metaphors, each of which attends to a different aspect of the informative object as a sign. The physical metaphor attends the signifier and its structural, logical, semantic, syntactic, and paradigmatic relations to other signifiers. Meaning is conceived as a matter of equivalencies between signifiers and is based on the notion that it is possible, on the basis of their functional attributes, to identify signifiers that can stand in for one another and to organize them in a way that controls information. From this perspective, "information" is a material, tangible, and rule-bound phenomenon that is external to consciousness and manifest in the informative object's role as sign-function in discourse. The physical metaphor attends the first-order language of texts.

In contrast, the cognitive metaphor attends the signified and its intellectual, affective, and meaningful relations to a user of information, and to other signifieds or concepts, especially concepts that a users brings to a reading of a text in order to interpret it. Meaning is conceived as a matter of differences, particularly in terms of the difference a text makes to its user and what difference is made by a user through the use of a text. At issue here is how the knowledge, experience, and reality of a user is changed through the use of a text. From this perspective, "information" is a non-material, intangible, creative phenomenon—essentially a matter of consciousness—and manifest in the content of an informative object. The cognitive metaphor attends the second-order language of texts.

In other words, the physical metaphor is concerned with the paradigmatic aspects of information while the cognitive metaphor is concerned with its syntagmatic aspects. This condition can be nicely illustrated by another of Barthes's analogies. Food is a sign system that displays a unity of paradigm and syntagm. The paradigm or system, as he calls it, is constituted by "a set of foodstuffs which have affinities or differences, within which one chooses a dish in view of a certain meaning: the types of entree, roast, or sweet." The syntagm is constituted by the "real sequence of dishes chosen during a meal: this is the menu."[42] Likewise information can be regarded as sign system in which, on one hand, there exists a set of texts within a discourse that have affinities or differences and, on the other hand, real sequences of texts actually chosen in an access situation. Indexing, or any other form of information control, brings to bear texts that exist in paradigmatic relation to one another, based on affinities or differences regarding their aboutness. Each text of the same paradigmatic order can in some way be substituted for any other text in the order. One chooses, i.e., retrieves, a text in view of a certain meaning that applies to all texts of a given order. Ordinarily, successful access is conceived as a matter of selecting texts whose paradigmatic order matches that of a query. However, a query can be complex, requiring the making of a menu such that it may be necessary to select texts from a variety of paradigmatic orders; topics, approaches, theories, and methods represent a few of the more likely types of texts that might be chosen.

Still, the selection of texts is only a part of access. How do the texts go together? How do the "tastes" of these texts combine? Let us consider the problem of a menu. Different types of dishes, and a wide variety of dishes from each type, may be selected to make a meal, but does that ensure the meal yields satisfaction? Does it deliver the desired experience? Does it appropriately signify the diner's wishes? What exactly does it signify about the diner? Does it signify something about the diner in the same way a raincoat signifies something about its wearer? These kinds of questions can be answered only by determining and investigating the syntagmatic relations between the objects selected—by the real sequence of objects chosen, from among texts or dishes, in a particular instance of choice. Regarding a meal, we can ask if the selection of a different particular dish from a type of dish, or a different type of dish altogether from among the types available, would make a difference to the meal and its experience, not just because there are differences, but because of the way these differences affect their unity as a combination?

The same condition holds true for texts. Two texts from the same paradigmatic order may not actually be entirely equivalent nor the difference trivial. Given a collection of texts retrieved in response to a query, the relevance of any one may depend as much on how it combines with the other texts selected as it does on any inherent attribute of the text itself and reminding us once again of the nature and consequences of intertextuality. The meaning of a single text in a discursive formation depends on its relations with all of the texts in that formation. Some combinations of text may yield only confusion, while other combinations yield an insight that could not have been anticipated. Clearly, both paradigmatic and syntagmatic aspects of information are related to relevance.

The "wrong" text can spoil a search for information in the same way the "wrong" adjective can spoil a sentence or the "wrong" dessert can spoil a meal. Let's suppose we've executed a search in an online database whose search strategy is "*A* and (*B* or *C*)", where *A*, *B*, and *C* are our search terms. We expect our results to include a set of documents that have been indexed with either term *B* or *C*, and this set is then examined for only those documents within it that have been indexed with term *A*. Let's further suppose that term *A* retrieves two documents, *x* and *y*. Both are "right" in the sense that the subject of both is accurately and adequately characterized by term *A*, but *x* is "wrong" in the sense that it will not contribute to a solution of the problem that motivated the search. If at least one of us has knowledge enough reject *x* as irrelevant, then we will be able to combine the documents indexed with *A*, *B*, and *C* in a way that will create a useful problem solving syntagm. Any other combination can cause problems. For example, if *x* is accepted, and *y*, which is relevant to our problem is rejected, or if *y*'s "rightness" influences our judgement of *x* and we accept both, or if *x*'s "wrongness" influences our judgment of *y* and we reject both, then we are likely to find ourselves in a state of confusion. Similarly, all of these problems can easily influence the reading and ultimately the relevance judgment of the texts retrieved with terms *B* and *C* that are associated with texts retrieved with term *A*. Thus, successful access to information, like a successful meal, depends not only on the individual items presented, but on their collective identity as an ensemble. It is entirely possible that a dinner of 'information' can be spoiled by the inclusion of the "wrong" dish.

Conclusion

I began this book with the assertion that "information" is an indeterminate theoretical object, not because its essence forever eludes understanding, but because it can be plausibly determined in so many—often incompatible— ways. The physical metaphor's positing of information-as-thing and the cognitive metaphor's positing of information-as-thought can now be regarded as different and differently articulated moments of the same phenomenon. The separation between these views is practical rather than categorical, a means of holding one aspect of information in suspension in order to examine another. Given our state of knowledge, this strategy is probably necessary and will be useful so long as we keep in mind that "information" is both thing and thought, and neither.

Information, like language, is always already everywhere. It is a given in things, processes, and minds, and it is a social institution—a well of shared knowledge and meanings conventionally attributed to reality. This attribution is signified through various of ways of knowing, including science, religion, ideology, and folklore. Like all cultural products, information assumes material form, and so its assumed identity as something natural and external to the self appears plausible, but its sources in culture reminds us of its reality as internal to the self. Information, manifest in material form, collectively exchanged, and individually internalized is then a foundational substance of community.

But just as not everyone shares the same language, not everyone shares the same information. Both conditions contribute to the creation and identity of different communities. The nature of and extent to which discourses constituted by information are shared allows us to distinguish equivalences and differences that mark different communities as being closely related or distantly separated, integrated or disintegrated, relevant or irrelevant to one another, and in consonance or conflict.

Also, like language, information manifests a latency. It exists only as a possibility until it is deliberately constituted by signs. In the case of language, the deployment of signs unites *langue* and *parole*. Language itself and on its own does not communicate until spoken, but without it there is nothing to say. Similarly, the signs that constitute information are informative objects. Informative objects, however, are special kinds of signs, already composed of other signs such as words, images, and sounds. These other signs are collectively and logically assembled and intended to communicate. As a whole, they constitute text which can be regarded as a special kind of signifier. Its counterpart, the signified, existing separately yet in unity with text, is content (Figure 11.4).

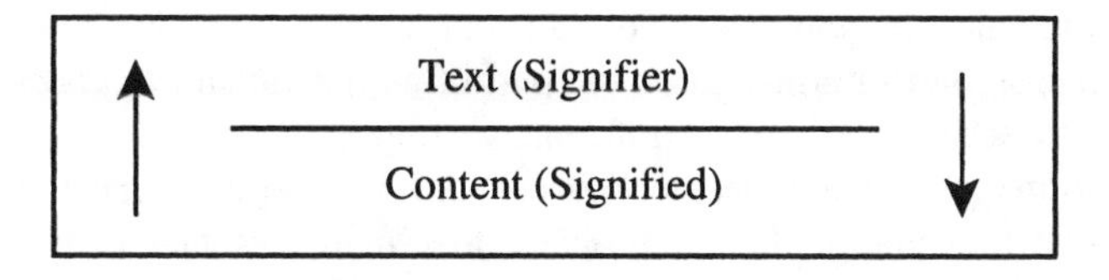

Figure 11.4 The Informative Object (Sign)

By means of the informative object, information is transformed into communication, a deliberate and intentional process known as discourse. From discourse in turn arise the discursive formations that constitute human reality. Like *langue*, information is shared and necessarily possesses a collective character. Like *parole*, the informative object is unique, the product of individual articulation and individual appropriation. Until information is given material form as an informative object, its reality remains latent. It exists as only one possibility, as Shannon might put it, among the many possibilities that could be communicated and play a role in a discursive formation.

However, in order for information to fulfill this role the informative object must be apprehended and appropriated by an interpreting subject, i.e., a user. Language does not come alive upon its being spoken but upon its being heard. Texts must be read for them to release their informative potential. At the moment of reading, the apparent dualities separating information-as-thing from information-as-thought melt into air. *Langue* can scarcely be separated from *parole*, nor *parole* from *langue.* They imply, require, and condition one another's existence. Buckland's insight that information-as-thing and information-as-thought can be interchanged through information-as-process, specifically the process of reading texts and becoming informed, ironically confirms these three apparently different phenomena are overdetermined by their equivalences and actually represent different moments of the same phenomenon, namely the phenomenon of information.

In the moment of information-as-process, when a text is read, Popper's Three Worlds, Dervin's Three Informations, and all of Buckland's distinctions regarding information collapse into one another. Content is apprehended and its information is released for appropriation by the reader. Text (thing) and content (cognitive structure) are reunited by the act of using the informative object. This act is composed of physical and cognitive, individual and social aspects, from which information is re-produced. The unity of signifier and signified, of text and content, is inherent in the informative object; but to complete the act of communication requires a twofold act of participation. First, a willful apprehension of information on the part of an individual must occur in order to access and use an informative object. Second, a cognitive transformation of that object must occur for it to enter into a meaningful and relevant relation to the problem that motivates and conditions participation. If this participation does not occur, then informing does not take place, and information falls back into the well of latency and existential possibility. Earlier in this text I posed the question, if information falls in a forest and no one hears it, is it information? The answer can only be *yes* and *no*.

The most intriguing aspect of bringing Saussure's work to bear on information as a theoretical object is that it accounts for the apparent separation of information into its physical, cognitive, and social aspects, even as it suggests a distinctly unitarian way of theoretically reconstituting these aspects into a meaningful whole. This outcome is especially important given Frohmann's cogent critique of information science as a discipline that tends to look away from information's status as a social phenomenon with the potential to inhibit rather than enhance human freedom.

On a more practical level, bringing semiotics to bear on information science allows us to understand how and why the discipline developed the two distinctly different research agendas observed by Ellis, one devoted to things and the other to people, when neither can operate without at least a tacit recognition of the presence of the other. Research based on the physical metaphor cannot entirely exclude people from its investigations of things. After all, the goal of this research is to provide users with access to information. Likewise, for the same reason, research based on the cognitive metaphor cannot entirely exclude things from its investigations of people. Each approach yields its insights precisely because it holds its apparent opposition with the other in suspension. The physical metaphor attends to the synchronous quality of information in its aspect as a signifier, and the cognitive metaphor attends to the diachronous quality of information in its aspect as a signified.

If we grant that information works like language, however, in ways congruent with both the physical and cognitive metaphor, then we can see how, through aboutness, representation, and relevance, the act of becoming informed involves the deliberate deployment and interpretation of signs. Both the objective attributes of texts and their interpretative apprehension and appropriation by subjects are located on the terrain of discourse. Aboutness, representation, and relevance are phenomena conditioned by social relations that constitute and are constituted by discourse. To paraphrase a well-known aphorism, this means that we can indeed *determine* the aboutness, representation, and relevance of information, but not exactly as we choose.

Information is clearly a matter of relations between thing and thought. However, thought and the discursive formations that sustain and reproduce it occur in a social context within which systems of information organization and retrieval, categories of aboutness, and a priori assumptions regarding relevance are, like language, socially constructed. No, these phenomena are not exclusively the products of power and expressions of a dominant hegemony, nor are rationality and choice subverted, precluded or denied. Neither is scientific investigation of the relations between thing and thought, and the means by which these relations constitute information, an impossible task, incapable of meaningfully advancing knowledge and solving of human problems.

Rather what we legitimately and conventionally determine to be information and informative is constrained and contingent on particular ideological, historical, and cultural conditions; and rationality and choice, while governed by rules that may indeed be objective, initially arise from the need to maintain power and sustain social relations, whether progressive or oppressive. Information is undoubtedly a social institution, but the control of this institution may not be evenly distributed across all segments of society. A need for information may be conditioned by what a dominant culture recognizes as legitimate and useful information, such that a need for any other kind of information is deemed illegitimate. At the same time, we must bear in mind that these constraints need not be imposed by official agents of authority, nor are they signs of a totalitarian polity.

Open societies, for example, manifest far fewer constraints than closed societies, but in fact sanctions may be subtle and culturally imposed. In capitalist democracies, information as a social institution is increasingly characterized by its commodity status. Its production, distribution, and use is determined by markets which are notorious for both their efficiency at distributing social goods and their discrimination against certain goods and needs that fail to find or create a large enough market. Thus, while the market for information is clearly based on economic rationality, fairness and egalitarianism are not automatic outcomes.

Systems of information collection, organization, and retrieval, systems of information access, and the conceptions of "information" as a theoretical object: all may appear to be natural, objective, and conforming to regularities discoverable and describable by science, when in fact they legitimize and confirm some needs as they displace and marginalize others. The key to this self-deception, as Frohmann implies, and the reason why we accept it, is that rational processes of discovering truths about and organizing information for access are grounded on unconscious and sometimes ideological premises. Distortions of rationality occur not in the process of investigating problems but in the moment of identifying and defining them. Now here is the truly intriguing point. Even if information is not a social institution of hegemony and power, even if relevance is not predetermined by what is it allowed to be, and even if information needs are not ideologically conditioned, it still stands to reason that none of these phenomena are free of situational contingency. We must still investigate the historical and cultural contingency of "information" as a theoretical object, and the influence of this contingency on whatever means we employ to understand the crucial theoretical dimensions of information: aboutness, representation, and relevance.

"Information" as a theoretical object is in an unenviable position. It must somehow embrace information as a material object, as an individual cognitive effect, and as a social institution. No wonder Fairthorne is so frustrated with the word. It is applied to signifier and signified, as well as to the cultural processes and conventions that condition the relations between the two and between the signs they constitute. As a result of the latter, it must also be accounted for as a commodity that exchanges for other commodities in both formal and informal markets. Information exists in a borderland between text and content, between consistency and contingency, between social convention and social conflict, between synchrony and diachrony, between message and meaning.

The project of information science is indeed a daunting one.

Endnotes

1. Ferdinand De Saussure, *Course in General Linguistics*, ed. Charles Bally and Albert Sechehaye, trans. Wade Baskin (New York: Philosophical Library, 1959).
2. Soren Brier, "Cybersemeotics: A New Disciplinary Development Applied to the Problems of Knowledge Organisation and Document Retrieval in Information Science," *Journal of Documentation* 52 no. 3 (September 1996); Gulten S. Wagner, *Public Libraries as Agents of Communication: A Semiotic Analysis* (Metuchen N.J.: Scarecrow Press, 1992); and Julian Warner, "Semiotics, Information Science, Documents and Computers," *Journal of Documentation* 46 no. 1 (March 1990).
3. Saussure, *Course*, 16.
4. Jere Paul Surber, *Culture and Critique: An Introduction to the Critical Discourses of Cultural Studies* (Boulder, Colo.: Westview Press, 1998), 164.
5. Roland Barthes. "The Kitchen of Meaning," *The Semiotic Challenge*, trans. Richard Howard (Berkeley, Calif.: University of California Press, 1994), 157-159.
6. Ibid., 157.
7. Ibid., 158.
8. Ibid., 158.
9. Michel Foucault, *The Archaeology of Knowledge*, trans. A.M. Sheridan Smith (New York: Pantheon Books, 1972), p. 31-39.
10. Saussure, *Course*, 8.
11. Ibid., 9, 11-13.
12. Ibid., 14.
13. Ibid.
14. Ibid., 15.
15. Ibid., 65.
16. Ibid., 67.
17. Ibid., 67-70.
18. Ibid., 74, 75.
19. Ibid., 79.
20. Ibid., 80.

21. Ibid., 102-103.
22. Ibid., 103.
23. Ibid.
24. Ibid., 107.
25. Ibid., 80.
26. Ibid., 108-109.
27. Ibid., 113. Italics in original.
28. Ibid.
29. Ibid., 14.
30. Ibid., 115.
31. Ibid.
32. Ibid., 118.
33. Ibid., 118-119.
34. Ibid., 120.
35. Roland Barthes, "The Semantics of Objects," in *The Semiotic Challenge*, trans. Richard Howard (Berkeley, Calif.: University of California Press, 1988, 1995), 179-190.
36. Ibid., 180.
37. Ibid., 181.
38. Ibid., 189.
39. Roland Barthes, *Elements of Semiology*, trans. Annette Lavers and Colin Smith (New York: Hill and Wang, 1967, 1973), p. 41.
40. Ibid., 41.
41. Ibid., 42.
42. Ibid., 63.

Bibliography

Aluri, Rao, D. Alasdair Kemp, and John J. Boll. "The Database." Chap. 2 in *Subject Analysis in Online Catalogs*. Englewood, Colo.: Libraries Unlimited, Inc., 1991.

———. "Language in Information Retrieval." Chap. 3 in *Subject Analysis in Online Catalogs*. Englewood, Colo.: Libraries Unlimited, Inc., 1991.

Anderson, James D. "Indexing and Classification: File Organization and Display for Information Retrieval." In *Indexing: The State of Our Knowledge and the State of Our Ignorance, Proceedings of the 20th Annual Meeting of the American Society of Indexers*, edited by Bella Hass Weinberg, 71-82. Medford, N.J.: Learned Information, Inc., 1989.

Asher, R. E., ed. *The Encyclopedia of Language and Linguistics*, Vol. 6. Oxford: Pergamon Press, 1994.

Barry, Carol L. "User-Defined Relevance Criteria: An Exploratory Study." *Journal of the American Society for Information Science* 45 (April 1994): 149-159.

Barthes, Roland. *Elements of Semiology*. Trans. Annette Lavers and Colin Smith. New York: Hill and Wang, 1967, 1973.

———. "The Kitchen of Meaning." *The Semiotic Challenge*. Trans. Richard Howard. Berkeley, Calif.: University of California Press, 1994.

———. "The Semantics of Objects." In *The Semiotic Challenge*. Trans. Richard Howard. Berkeley, Calif.: University of California Press, 1994.

Beghtol, Clare. "Bibliographic Classification Theory and Text Linguistics: Aboutness Analysis, Intertextuality and the Cognitive Act of Classifying Documents." *Journal of Documentation* 42 (June 1986): 84-113.

Belkin, N. J. "The Cognitive Viewpoint in Information Science." *Journal of Information Science* 16 (1990): 11-15.

———. "Progress in Documentation: Information Concepts for Information Science." *Journal of Documentation* 34 (March 1978): 55-85.

Belkin, N. J., H. M. Brooks, and P. J. Daniels. "Knowledge Elicitation Using

Discourse Analysis." *International Journal of Man-Machine Studies* 27 (1987): 127-144.

Belkin, N. J., R. N. Oddy, and H. M. Brooks. "ASK for Information Retrieval: Part I. Background and Theory." *Journal of Documentation* 38 (June 1982): 61-71.

———. "ASK for Information Retrieval: Part II. Results of a Design Study." *Journal of Documentation* 38 (September 1982): 145-164.

Belkin, N. J., and S. E. Robertson. "Information Science and the Phenomenon of Information." *Journal of the American Society for Information Science* 27 (July-August 1976): 197-204.

Belkin, N. J., T. Seeger, and G. Wersig. "Distributed Expert Problem Treatment as a Model for Information System Analysis and Design." *Journal of Information Science* 5 (1983): 153-167.

Beniger, James R. "Information and Communication: The New Convergence." *Communication Research* 15 (April 1988): 198-218.

Berger, Peter L., and Thomas Luckmann. *The Social Construction of Reality.* Garden City, N.Y.: Doubleday Anchor Books, 1967.

Bingi, R., Deepak Khazanchi, and Surya B. Yadav. "A Framework for the Comparative Analysis and Evaluation of Knowledge Representation Schemes." *Information Processing & Management* 31, no. 2 (1995): 233-247.

Bookstein, A. "Explanations of Bibliometric Laws." *Collection Management* 3 (Summer/Fall 1979): 151-162.

Borgman, Christine L. "Bibliometrics and Scholarly Communication: Editor's Introduction." *Communication Research* 16 (October 1989): 583-599.

Borko, Harold. "Information Science: What Is It?" In *Key Papers in Information Science*, edited by A. W. Elias, 1-3. Washington, D.C.: American Society for Information Science, 1971.

———. "Toward a Theory of Indexing." *Information Processing & Management* 13, no. 6 (1977): 355-365.

Borko, Harold, and C. L. Bernier. *Indexing Concepts and Methods*. Newe York: Academic Press, 1978.

Bouissac, Paul, ed. *Encyclopedia of Semiotics*. Oxford: Oxford University Press, 1998.

Bradford, S. C. "Sources of Information on Specific Subjects." *Engineering* 137 (1935): 85-86.

Brier, Soren. "Cybersemiotics: A New Interdisciplinary Development Applied to the Problems of Knowledge Organization and Document Retrieval in Information Science." *Journal of Documentation* 52 no. 3 (September 1996): 296-344.

———. "A Philosophy of Science Perspective—On the Idea of a Unifying Information Science." In *Conceptions of Library and Information Science: Historical, Empirical and Theoretical Perspectives*, edited by Pertti Vakkari and Blaise Cronin, 97-108. London: Taylor Graham, 1992.

Brookes, Bertram C. "The Foundations of Information Science: Part I. Philosophical Aspects." *Journal of Information Science* 2 (1980): 125-133.

———. "Robert Fairthorne and the Scope of Information Science." *Journal of*

Documentation 30 (June 1974): 139-152.

Brooks, H.M. "Expert System and Intelligent Information Retrieval." *Information Processing & Management* 23, no. 4 (1987): 367-382.

Buckland, Michael K. *Information and Information Systems.* New York: Greenwood Press, 1991.

———. "Information as Thing." *Journal of the American Society for Information Science* 42 (June 1991): 351-360.

Budd, John M. "An Epistemological Foundation for Library and Information Science." *Library Quarterly* 65, no. 3 (1995): 295-318.

———. "User-Centered Thinking: Lessons from Reader-Centered Theory." *RQ* 34 (Summer 1995): 487-496.

Bush, Vannevar. "As We May Think," *Atlantic Monthly* (July 1945): 101-108.

Cleveland, Donald B., and Ana D. Cleveland. *Introduction to Indexing and Abstracting*. Littleton, Colo.: Libraries Unlimited, 1983.

Cleverdon, Cyril W. "The Cranfield Tests of Index Language Devices." *Aslib Proceedings* 19 (June 1967): 173-194.

———. "On the Inverse Relationship of Recall and Precision." *Journal of Documentation* 28 (September 1972): 195-201.

———. "Review of the Origins and Development of Research." *Aslib Proceedings* 22 (November): 538-549.

Cole, Charles. "Operationalizing the Notion of Information as a Subjective Construct." *Journal of the American Society for Information Science* 45 (August 1994): 465-476.

Cool, Colleen. "Information Retrieval as Symbolic Interaction: Examples from Humanities Scholars." In *Proceedings of the 56th Annual Meeting of the American Society for Information Science,* edited by Susan Bonzi, 274-277. Medford, N.J.: Learned Information, Inc., for American Society for Information Science, 1993.

Cooper, W. S. "A Definition of Relevance for Information Retrieval." *Information Storage and Retrieval* 7 (1971): 19-37.

Croft, W. Bruce. "Approaches to Intelligent Information Retrieval." *Information Processing & Management* 23, no. 4 (1987): 249-254.

———. "Automatic Indexing." In *Indexing: The State of Our Knowledge and the State of Our Ignorance, Proceedings of the 20th Annual Meeting of the American Society of Indexers,* edited by Bella Hass Weinberg, 86-100. Medford, N.J.: Learned Information, Inc., 1989.

Croft, W. Bruce, and R. H. Thompson. "I³R: A New Approach to the Design of Document Retrieval Systems." *Journal of the American Society for Information Science* 38 (November 1987): 389-404.

Cuadra, Carlos A. "On-line Systems: Promise and Pitfalls." *Journal of the American Society for Information Science* 22 (March-April 1971): 107-114.

Cuadra, Carlos A., and Robert V. Katter. "Opening the Black Box of 'Relevance.'" *Journal of Documentation* 23 (December 1967): 291-303.

Davenport, Elisabeth. "Information Science Observed: New Media and Productivity in a Group of UK Practitioners." *Journal of Documentation* 50

(December 1994): 291-315.

Davis, Charles H., and James E. Rush. *Guide to Information Science*. Westport, Conn.: Greenwood Press, 1979.

DeMay, M. "The Cognitive Viewpoint: Its Development and Its Scope." In *International Workshop on the Cognitive Viewpoint*. Ghent: University of Ghent, Belgium, 1977.

Dervin, Brenda. "Useful Theory for Librarianship: Communication, Not Information." *Drexel Library Quarterly* 13, no. 3 (1977): 16-32.

Ellis, David. "The Dilemma of Measurement in Information Retrieval Research." *Journal of the American Society for Information Science* 47 (January 1996): 23-36.

———. *New Horizons in Information Retrieval*. London: The Library Association, 1990.

———."The Physical and Cognitive Paradigms in Information Retrieval Research." *Journal of Documentation* 48 (March 1992): 45-64.

Ellis, David, Deborah Cox, and Katherine Hall. "A Comparison of the Information Seeking Patterns of Researchers in the Physical and Social Sciences." *Journal of Documentation* 49 (December 1993): 356-369.

Eulau, Heinz. *The Behavioral Persuasion in Politics*. New York: Random House, 1963.

Faibisoff, Sylvia G., and Donald P. Ely. "Information and Information Needs." *Information Reports and Bibliographies* 5, no. 5 (1976): 2-16.

Fairthorne, Robert A. "Empirical Hyperbolic Distributions (Bradford—Zipf—Mandlebrot) for Bibliometric Description and Prediction." *Journal of Documentation* 25 (December 1969): 319-343.

———. "Information: One Label, Several Bottles." In *Perspectives in Information Science: Proceedings of the NATO Advanced Study Institute on Perspectives in Information Science,* edited by Anthony Debons and W. J. Cameron, 65-73. Leyden, The Netherlands: Noordhoff, 1975.

Farradane, J. "The Nature of Information." *Journal of Information Science* 1 (1979): 13-17.

Fidel, Raya. "User-Centered Indexing." *Journal of the American Society for Information Science* 45 (September 1994): 572-576.

Foskett, Antony C. *The Subject Approach to Information*. 4th ed. London: Bingley, 1982.

Foucault, Michel. *The Archaeology of Knowledge*. Trans. A.M. Sheridan Smith. New York: Pantheon Books, 1972.

Froehlich, Thomas J. "Relevance Reconsidered—Towards an Agenda for the 21st Century: Introduction to Special Topic Issue on Relevance Research." *Journal of the American Society for Information Science* 45 (April 1994): 124-134.

Frohmann, Bernd. "Communication Technologies and the Politics of Postmodern Information Science." *Canadian Journal of Information and Library Science* 19 (July 1994): 1-22.

———. "Discourse Analysis as a Research Method in Library and Information Science." *Library and Information Science Research* 16 (1994): 119-138.

———. "The Power of Images: A Discourse Analysis of the Cognitive Viewpoint." *Journal of Documentation* 48 (December 1992): 365-386.

———. "Rules of Indexing: A Critique of Mentalism in Information Retrieval Theory." *Journal of Documentation* 46 (June 1990): 81-101.

Gapen, D. Kaye, and Sigrid P. Milner. "Obsolescence." *Library Trends* 30 (Summer 1981): 107-124.

Garfield, Eugene. "Citation Indexes for Science: A New Dimension in Documentation Through Association of Ideas." *Science* 122 (July 1955): 108-111.

Goffman, William. "A General Theory of Communication." In *Introduction to Information Science*, edited by Tefko Seracevic, 726-747. New York: Bowker Co., 1970.

———. "Information Science: Discipline or Disappearance." *Aslib Proceedings* 22 (December 1970): 589-596.

Green, Rebecca. "The Profession's Models of Information: A Cognitive Linguistic Analysis." *Journal of Documentation* 47 (June 1991): 130-148.

Griffith, Belver C. "Understanding Science: Studies of Communication and Information." *Communication Research* 16 (October 1989): 600-614.

Griffiths, José-Marie, and Donald W. King. "Libraries: The Undiscovered National Resource." In *The Value and Impact of Information*, edited by Mary Feeney and Maureen Grieves, 79-116. London: Bowker Saur, 1994.

Gutting, Gary, ed. *Paradigms and Revolutions: Applications of Thomas Kuhn's Philosophy of Science*. Notre Dame, Ind.: University of Notre Dame Press, 1980.

Harris, Michael H., and Stan A. Hannah. *Into the Future: The Foundations of Library and Information Services in the Post-Industrial Era*. Norwood, N.J.: Ablex Publishing Co., 1993.

Harter, Stephen P. "The Impact of Electronic Journals on Scholarly Communication: A Citation Analysis." *The Public-Access Computer Systems Review* 7, no. 5 (1996). At http://info.lib.uh.edu/pr/v7/n5/hart7n5.html (accessed 10 February 2000).

———. "Psychological Relevance and Information Science." *Journal of the American Society for Information Science* 43 (October 1992): 602-615.

———. "Variations in relevance Assessments and the Measurement of Retrieval Effectiveness." *Journal of the American Society for Information Science* 47 (January 1996): 37-49.

Hjørland, Birger. "The Concept of 'Subject' in Information Science." *Journal of Documentation* 48 (June 1992): 172-200.

Hjørland, Birger, and Hanne Albrechtsen. "Toward a New Horizon in Information Science: Domain-Analysis." *Journal of the American Society for Information Science* 46 (July 1995): 400-425.

Hoel, Ivar A. L. "Information Science and Hermeneutics—Should Information Science be Interpreted as a Historical and Humanistic Science?" In *Conceptions of Library and Information Science: Historical, Empirical and Theoretical Perspectives*, edited by Pertti Vakkari and Blaise Cronin, 69-81.

London: Taylor Graham, 1992.

Hutchins, W. J. "The Concept of 'Aboutness' in Subject Indexing." *Aslib Proceedings* 30 (May 1978): 172-181.

Ingwersen, Peter. "Cognitive Perspectives of Information Retrieval Interaction: Elements of a Cognitive IR Theory." *Journal of Documentation* 52 (March 1996): 3-50.

———. *Information Retrieval Interaction.* London: Taylor Graham, 1992.

Jakobson, R. "Linguistics and Poetics." *Style and Language*. Ed. T. A. Sebeok. Cambridge, Mass.: MIT Press, 1960.

Järvelin, Kalervo, and Pertti Vakkari. "The Evolution of Library and Information Science 1965-1985: A Content Analysis of Journal Articles." In *Conceptions of Library and Information Science: Historical, Empirical and Theoretical Perspectives*, edited by Pertti Vakkari and Blaise Cronin, 109-125. London: Taylor Graham, 1992.

Johann-Glock, Hans. "Words and Things." *Prospect* April 1999 http://www.prospect-magazine.co.uk/highlights/words_things_apr99/index.html (accessed 10 February 2000).

Johanson, Graeme. "Information, Knowledge and Research." *Journal of Information Science* 23, no. 2 (1997): 103-109.

Jones, Karen Sparck. "The Cranfield Tests." In *Information Retrieval Experiment*, edited by Karen Sparck Jones, 256-284. London: Butterworths, 1981

Kaplan, Abraham. *The Conduct of Inquiry: Methodology for the Behavioral Sciences*. Scranton, Pa.: Chandler Publishing Company, 1964.

Kilgour, Frederick G. "Computers for University Libraries: Introduction." *Drexel Library Quarterly* 4, no. 3 (1968): 155-156.

———. "University Libraries and Computation." *Drexel Library Quarterly* 4, no. 3 (1968): 157-176.

Kuhlthau, Carol C. "A Principle of Uncertainty for Information Seeking." *Journal of Documentation* 49 (December 1993): 339-355.

Kuhn, Thomas. *The Structure of Scientific Revolutions*. 2nd ed. Chicago: University of Chicago Press, 1970.

Lakatos, Imre. *The Methodology of Scientific Research Programmes: Philosophical Papers, Vol. 1*. Ed. John Worrall and Gregory Currie. New York: Cambridge University Press, 1978.

Lakoff, George. "The Importance of Categorization." Chap. 1 in *Women, Fire, and Dangerous Things: What Categories Reveal About the Mind.* Chicago: University of Chicago Press, 1987.

Lakoff, George and Mark Johnson. *Metaphors We Live By*. Chicago: University of Chicago Press, 1980.

Lancaster, F. Wilfred. "Automatic Indexing, Automatic Abstracting and Related Procedures." Chap. 14 in *Indexing and Abstracting in Theory and Practice*. London: Library Association Publishing, 1991.

———. *Characteristics, Testing and Evaluation*, 2d ed. New York: John Wiley & Sons, 1979.

———. *The Measurement and Evaluation of Library Services.* Washington, D.C.: Information Resources Press, 1977.

———. *Vocabulary Control for Information Retrieval.* 2nd ed. Arlington, Va.: Information Resources Press, 1986.

Langridge, Derek. "Classifying Knowledge." In *Knowledge and Communication: Essays on the Information Chain,* edited by A. J. Meadows, 1-18. London: Library Association Publishing, 1991.

Lilley, Dorothy B., and Ronald W. Trice. *A History of Information Science, 1945-1985.* San Diego, Calif.: Academic Press, 1989.

Lincoln, Yvonne S. and Egon G. Guba. *Naturalistic Inquiry.* Newbury Park, Calif.: Sage Publications, 1985.

Losee, Robert M. "A Discipline Independent Definition of Information." *Journal of the American Society for Information Science* 48 (March 1997): 254-269.

Lotka, A. J. "The Frequency Distribution of Scientific Productivity." *Journal of the Washington Academy of Sciences* 16 (1926): 317-323.

Luhn, H. P. "The Automatic Creation of Literature Abstracts." *IBM Journal of Research and Development* 2 (April 1958): 159-165.

Machlup, Fritz, and Una Mansfield. "Cultural Diversity in Studies of Information." In *The Study of Information: Interdisciplinary Messages,* edited by Fritz Machlup and Una Mansfield, 3-56. New York: Wiley, 1983.

Maron, M. E. "Automatic Indexing: An Experimental Inquiry." *Journal of the Association for Computing Machinery* 8 (1961): 404-417.

———. "On Indexing, Retrieval and the Meaning of About." *Journal of the American Society for Information Science* 28 (January 1977): 38-43.

Masterman, Margaret. "The Nature of a Pardigm." In *Criticism and the Growth of Knowledge.* Ed. I Lakatos and A. Musgrave. Cambridge: Cambridge University Press, 1970.

Maze, Susan, David Moxley, and Donna J. Smith. *Authoritative Guide to Web Search Engines.* New York: Neal-Schuman, 1997.

McCain, Garvin, and Erwin M. Segal. *The Game of Science.* 2nd ed. Monterey, Calif.: Brooks/Cole Publishing Company, 1973.

Menou, Michel J. "The Impact of Information—I. Toward a Research Agenda for its Definition and Measurement." *Information Processing & Management* 31, no. 4 (1995): 455-477.

———. "The Impact of Information—II. Concepts of Information and Its Value." *Information Processing & Management* 31, no. 4 (1995): 479-490.

Miksa, Francis L. "Library and Information Science: Two Paradigms." In *Conceptions of Library and Information Science: Historical, Empirical and Theoretical Perspectives,* edited by Pertti Vakkari and Blaise Cronin, 229-252. London: Taylor Graham, 1992.

Milstead, Jessica L. "Needs for Research in Indexing." *Journal of the American Society for Information Science* 45 (September 1994): 577-582.

Morris, Ruth C. T. "Toward a User-Centered Information Service." *Journal of the American Society for Information Science* 45 (January 1994): 20-30.

Mullins, Nicholas C. *Theories and Theory Groups in Contemporary American*

Sociology. New York: Harper & Row, 1973.

Neill, S. D. "The Dilemma of the Subjective in Information Organisation and Retrieval." *Journal of Documentation* 43 (September 1987): 193-211.

———. *Dilemmas in the Study of Information: Exploring the Boundaries of Information Science*. New York: Greenwood Press, 1992.

Nitecki, Danuta A. "Conceptual Models of Libraries Held by Faculty, Administrators, and Librarians: An Exploration of Communications in the Chronicle of Higher Education." *Journal of Documentation* 49 (September 1993): 255-277.

Nossiter, T. J., A. H. Hanson, and Stein Rokkan, ed. *Imagination and Precision in the Social Sciences*. London: Faber and Faber, 1972.

O'Connor, Brian C. *Explorations in Indexing and Abstracting: Pointing, Virtue, and Power*. Englewood, Colo.: Libraries Unlimited, 1996.

O'Connor, Daniel O., and Henry Voos. "Empirical Laws, Theory Construction and Bibliometrics." *Library Trends* 30 (Summer 1981): 9-20.

Oddy, R. N. "Information Retrieval Through Man-Machine Dialogue." *Journal of Documentation* 33 (March 1977): 1-14.

Olsgaard, John N. Introduction to *Principles and Applications of Information Science for Library Professionals,* edited by John N. Olsgaard. Chicago: American Library Association, 1989.

Otten, Klaus, and Anthony Debons. "Towards a Metascience of Information: Informatology." *Journal of the American Society for Information Science* 21 (January-February 1970): 89-94.

Paisley, William. "Information Science as a Multidiscipline." In *Information Science: The Interdisciplinary Context*, edited by J. Michael Pemberton and Ann E. Prentice, 3-24. New York: Neal-Schuman Publishers, Inc., 1990.

Park, Taemin Kim. "The Nature of Relevance in Information Retrieval: An Empirical Study." *Library Quarterly* 63, no. 3 (1993): 318-351.

———. "Toward a Theory of User-Based Relevance: A Call for a New Paradigm of Inquiry." *Journal of the American Society for Information Science* 45 (April 1994): 135-141.

Pfaffenberger, Bryan. *Democratizing Information: Online Databases and the Rise of End-User Searching*. Boston: G. K. Hall, 1990.

Popper, Karl. *The Logic of Scientific Discovery*. New York: Basic Books, 1959.

Pratt, Allan D. "The Information of the Image: A Model of the Communications Process." *Libri* 27, no. 3 (1977): 204-220.

Pritchard, Alan. "Statistical Bibliography or Bibliometrics?" *Journal of Documentation* 25 (December 1969): 343-357.

Radford, Gary P. "Positivism, Foucault, and the Fantasia of the Library: Conceptions of Knowledge, and the Modern Library Experience." *Library Quarterly* 62, no. 4 (1992): 408-424.

Rayward, W. Boyd. "The History and Historiography of Information Science: Some Reflections." *Information Processing & Management* 32, no. 1 (1996): 3-17.

Rigney, Daniel. *The Metaphorical Society: An Invitation to Social Theory*.

Lanham, Md.: Rowman & Littlefield Publishers, Inc., 2000.

Roberts, Norman. "Social Considerations Towards a Definition of Information Science." *Journal of Documentation* 32 (December 1976): 249-257.

Robertson, S. E. "The Methodology of Information Retrieval Experiment." In *Information Retrieval Experiment*, edited by Karen Sparck Jones, 9-31. London: Butterworths, 1981.

Robertson, S. E., and M. M. Hancock-Beaulieu. "On the Evaluation of IR Systems." *Information Processing & Management* 28, no. 4 (1992): 457-466.

Rosenbaum, Howard. "Information Use Environments and Structuration: Towards an Integration of Taylor and Giddens." In *Proceedings of the 56th Annual Meeting of the American Society for Information Science,* edited by Susan Bonzi, 235-245. Medford, N.J.: Learned Information, Inc., for American Society for Information Science, 1993.

Ruben, Brent D. "The Communication-Information Relationship in System-Theoretic Perspective." *Journal of the American Society for Information Science* 43 (January 1992): 15-27.

Salton, Gerard. "The Smart Environment for Retrieval System Evaluation—Advantages and Problem Areas." In *Information Retrieval Experiment*, edited by Karen Sparck Jones, 316-329. London: Butterworths, 1981.

———. "The State of Retrieval System Evaluation." *Information Processing & Management* 28, no. 4 (1992): 441-449.

Samuelson, Kjel. "Information Models and Theories—a Synthesizing Approach." In *Information Science: Search for Identity: Proceedings of the 1972 NATO Advanced Study Institute in Information Science*, edited by Anthony Debons, 47-67. New York: M. Dekker, 1974.

Sandison, Alexander. "Thinking About Citation Analysis." *Journal of Documentation* 45 (March 1989): 59-64.

Saracevic, Tefko. "Relevance: A Review of and a Framework for the Thinking on the Notion in Information Science." *Journal of the American Society for Information Science* 26 (November-December 1975): 321-343.

Saracevic, Tefko, and Paul Kantor. "A Study of Information Seeking and Retrieving: II. Questions, and Effectiveness." *Journal of the American Society for Information Science* 39 (May 1988): 177-196.

———. "A Study of Information Seeking and Retrieving: III. Searchers, Searches, and Overlap." *Journal of the American Society for Information Science* 39 (May 1988): 197-216.

Saracevic, Tefko, Paul Kantor, Alice Y. Chamis, and Donna Trivison. "A Study of Information Seeking and Retrieving: I. Background and Methodology." *Journal of the American Society for Information Science* 39 (May 1988): 161-176.

Saussure, Ferdinand de. *Course in General Linguistics*. Ed. Charles Bally and Albert Sechehaye. Trans. Wade Baskin. New York: Philosophical Library, 1959.

Savolainen, Reijo. "The Sense-Making Theory: Reviewing the Interests of a User-Centered Approach to Information Seeking and Use." *Information Processing*

& Management 29 (1993): 13-28.

Schrader, Alvin M. "In Search of a Name: Information Science and Its Conceptual Antecedents." *Library & Information Science Research* 6 (1984): 227-271.

Shera, Jesse. *Foundations of Education for Librarianship*. New York: Becker and Hayes, Inc. John Wiley & Sons, 1972

———. *Knowing Books and Men; Knowing Computers, Too*. Littleton, Colo: Libraries Unlimited, Inc., 1973.

———. *Libraries and the Organization of Knowledge*. Ed. and with an introduction by D. J. Foskett. Hamden, Conn.: Anchon Books, 1965.

Siever, Mary Ellen, Emma Jean McKinin, E. Diane Johnson, and John C. Reid. "Beyond Relevance—Characteristics of Key Papers for Clinicians: An Exploratory Study in an Academic Setting." *Bulletin of the Medical Library Association* 84 (July 1996): 351-358.

Smith, Linda C. "Citation Analysis." *Library Trends* 30 (Summer 1981): 83-106.

Soergel, Dagobert. "Indexing and Retrieval Performance: The Logical Evidence." *Journal of the American Society for Information Science* 45 (September 1994): 589-599.

Stinchombe, Arthur. *Constructing Social Theories*. New York: Harcourt, Brace & World, 1968.

Stove, D. C. *Popper and After: Four Modern Irrationalists*. New York: Pergamon Press, 1982.

Surber, Jere Paul. *Culture and Critique: An Introduction to the Critical Discourses of Cultural Studies*. Boulder, Col.: Westview Press, 1998.

Svenonius, Elaine, and Helen F. Schmierer. "Current Issues in the Subject Control of Information." *Library Quarterly* 47, no. 3 (1977): 326-346.

Swanson, Don R. "Subjective Versus Objective Relevance in Bibliographic Retrieval Systems." *Library Quarterly* 56, no. 4 (1986): 389-398.

Swift, D. F., V. Winn, and D. Bramer. "'Aboutness' as a Strategy for Retrieval in the Social Sciences." *Aslib Proceedings* 30 (May 1978): 182-187.

Tague-Sutcliffe, Jean M. "Some Perspectives on the Evaluation of Information Retrieval Systems." *Journal of the American Society for Information Science* 47 (January 1996): 1-3.

Taylor, Robert S. "Question-Negotiation and Information Seeking in Libraries." *College & Research Libraries* 29 (May 1968): 178-194.

———. *Value-Added Processes in Information Systems*. Norwood, N.J.: Ablex Publishing Co., 1986.

van Rijsbergen, C. J. "The Science of Information Retrieval: Its Methodology and Logic." in *Conference Informatienvetenschap in Nederland*. Hagg, The Netherlands: Rabin, 1990.

Vickery, B.C. "Knowledge Representation: A Brief Review." *Journal of Documentation* 42 (September 1986): 145-159.

Wagner, Gulten S. *Public Libraries as Agents of Communication*. Metuchen, N.J.: Scarecrow Press, 1992.

Wall, Eugene. "Symbiotic Development of Thesauri and Information Systems: A Case History." *Journal of the American Society for Information Science* 26

(March-April 1975): 71-79.
Wallace, Danny P. "Bibliometrics and Citation Analysis." In *Principles and Applications of Information Science for Library Professionals*, edited by John Olsgaard, 10-26. Chicago: American Library Association, 1989.
Warner, Julian. "Semiotics, Information Science, Documents and Computers." *Journal of Documentation* 46 (March 1990): 16-32.
Weaver, Warren. "The Mathematics of Communication." *Scientific American* 181 (July 1949): 11-15.
———. "Recent Contributions to the Mathematical Theory of Communication." In *The Mathematical Theory of Communication* by Claude E. Shannon and Warren Weaver, 93-117. Urbana, Ill.: University of Illinois Press, 1949.
Wellisch, H. H. *Indexing From A to Z*. New York: H. W. Wilson, 1991.
Wersig, G. "Information Science: The Study of Postmodern Knowledge Usage." *Information Processing & Management* 29, no. 2 (1993): 229-239.
———. "Information Science and Theory: A Weaver Bird's Perspective." In *Conceptions of Library and Information Science: Historical, Empirical and Theoretical Perspectives*, edited by Pertti Vakkari and Blaise Cronin, 201-217. London: Taylor Graham, 1992.
Wersig, G., and Ulrich Neveling. "The Phenomena of Interest to Information Science." *The Information Scientist* 9 (December 1975): 127-140.
White, Howard D., and Katherine W. McCain. "Visualizing a Discipline: An Author Co-Citation Analysis of Information Science, 1972-1995." *Journal of the American Society for Information Science* 49 (April 1998): 327-355.
Williams, Raymond. *Keywords: A Vocabulary of Culture and Society*. Rev. ed. New York: Oxford University Press, 1983.
Wilson, T. D. "On User Studies and Information Needs." *Journal of Documentation* 37 (March 1981): 3-15.
Yiexiao, Zhang. "Definitions and Sciences of Information." *Information Processing & Management* 24, no. 4 (1988): 479-491.
Yoon, Lanju Lee. "The Performance of Cited References as an Approach to Information Retrieval." *Journal of the American Society for Information Science* 45 (June 1994): 287-299.
Zipf, George K. *Human Behavior and the Principle of Least Effort*. Cambridge, Mass.: Addison-Wesley Press, 1949.

Index

About the Author

Douglas Raber (B.A., M.A., Indiana University; M.A.L.S., Northern Illinois University; Ph.D., Indiana University) has thirteen years of professional experience in academic and public libraries. He is currently an associate professor at the School of Information Sciences, University of Tennessee, Knoxville. His research and teaching interests include American national information and telecommunications policy, public libraries and the social construction of intellectual freedom. He is a member of the Tennessee Advisory Council on Libraries and serves on the editorial board of *Reference and User Services Quarterly*. He is the author of *Librarianship and Legitimacy: The Ideology of the Public Library Inquiry* (1997).